T0201235

Dealing with Aging Process Facilities and Infrastructure

This book is one in a series of process safety guidelines and concept books published by the Center for Chemical Process Safety (CCPS). Please go to *www.wiley.com/go/ccps* for a full list of titles in this series.

It is sincerely hoped that the information presented in this document will lead to an even more impressive safety record for the entire industry. However, the American Institute of Chemical Engineers, its consultants, the CCPS Technical Steering Committee and Subcommittee members, their employers, their employers' officers and directors, and AcuTech Consulting Group, Inc., and its employees do not warrant or represent, expressly or by implication, the correctness or accuracy of the content of the information presented in this document. As between (1) American Institute of Chemical Engineers, its consultants, CCPS Technical Steering Committee and Subcommittee members, their employers, their employers' officers and directors, and AcuTech Consulting Group, Inc., and its employees and (2) the user of this document, the user accepts any legal liability or responsibility whatsoever for the consequences of its use or misuse.

Dealing with Aging Process Facilities and Infrastructure

CENTER FOR CHEMICAL PROCESS SAFETY
of the
AMERICAN INSTITUTE OF CHEMICAL ENGINEERS
New York, NY

WILEY

This edition first published 2018

© 2018 the American Institute of Chemical Engineers

All rights reserved. No part of this publication may be reproduced, stored in a retrieval system, or transmitted, in any form or by any means, electronic, mechanical, photocopying, recording or otherwise, except as permitted by law. Advice on how to obtain permission to reuse material from this title is available at http://www.wiley.com/go/permissions.

The right of CCPS to be identified as the author of this work has been asserted in accordance with law.

Registered Office
John Wiley & Sons, Inc., 111 River Street, Hoboken, NJ 07030, USA

Editorial Office
111 River Street, Hoboken, NJ 07030, USA

For details of our global editorial offices, customer services, and more information about Wiley products visit us at www.wiley.com.

Wiley also publishes its books in a variety of electronic formats and by print-on-demand. Some content that appears in standard print versions of this book may not be available in other formats.

Limit of Liability/Disclaimer of Warranty
While the publisher and authors have used their best efforts in preparing this work, they make no representations or warranties with respect to the accuracy or completeness of the contents of this work and specifically disclaim all warranties, including without limitation any implied warranties of merchantability or fitness for a particular purpose. No warranty may be created or extended by sales representatives, written sales materials or promotional statements for this work. The fact that an organization, website, or product is referred to in this work as a citation and/or potential source of further information does not mean that the publisher and authors endorse the information or services the organization, website, or product may provide or recommendations it may make. This work is sold with the understanding that the publisher is not engaged in rendering professional services. The advice and strategies contained herein may not be suitable for your situation. You should consult with a specialist where appropriate. Further, readers should be aware that websites listed in this work may have changed or disappeared between when this work was written and when it is read. Neither the publisher nor authors shall be liable for any loss of profit or any other commercial damages, including but not limited to special, incidental, consequential, or other damages.

Library of Congress Cataloging-in-Publication Data:

Names: American Institute of Chemical Engineers. Center for Chemical Process Safety, author.
 Title: Dealing with aging process facilities and infrastructure / Center for Chemical Process
Safety of the American Institute of Chemical Engineers. Description: New York, NY :
    Wiley, 2018. | Includes bibliographical references and index. |
Identifiers: LCCN 2018010560 (print) | LCCN 2018012416 (ebook) |
 ISBN 9781119430766 (pdf) | ISBN 9781119430759 (epub) | ISBN 9781119430834 (cloth)
Subjects: LCSH: Chemical plants--Maintenance and repair. | Chemical
   plants--Equipment and supplies--Deterioration. | Service life (Engineering)
Classification: LCC TP155.5 (ebook) | LCC TP155.5 .D425 2018 (print) | DDC 660--dc23
LC record available at https://lccn.loc.gov/2018010560

Cover images: Courtesy of CCPS
Cover design by Wiley

Printed in the United States of America

10 9 8 7 6 5 4 3 2 1

TABLE OF CONTENTS

LIST OF TABLES

LIST OF FIGURES

ACKNOWLEDGMENTS

The American Institute of Chemical Engineers (AIChE) and the Center for Chemical Process Safety (CCPS) express their appreciation and gratitude to all members of the Aging Process Facilities and Infrastructure Subcommittee for their generous efforts in the development and preparation of this important concept book. CCPS also wishes to thank the subcommittee members' respective companies for supporting their involvement in this project.

We appreciate the involvement and writing contributions of Brian Kelly and Terry White. Special thanks are extended to the team of technical writers from ioMosaic Corporation who coordinated inputs and developed the manuscript. The ioMosaic team consisted of Elena Prats, Peter Stickles and Kathy Anderson.

The members of the CCPS project subcommittee were:

Eric Freiburger	Praxair, subcommittee chair
Brian Kelly	CCPS staff consultant
Laura Bellman	Covestro
Larry Bowler	SABIC
Bill Callaghan	Nova Chemicals
Derin Adebekun	Air Products
Susan Lubell	Nexen Energy
Bennie Barnes	Pacific Gas and Electric
Jonas Duarte	Chemtura
Reyyan Koc	ExxonMobil Chemical
John Murphy	CCPS emeritus
Jatin Shah	BakerRisk
Ken Tague	Archer Daniels Midland
Sudhir Phakey	Linde
Nancy Faulk	Siemens Energy
Tom Sandbrook	Chemours
Robb Van Sickle	Flint Hills Resources
Terry White	Pacific Gas and Electric
Bob Wasileski	formerly Nova Chemicals

All CCPS books are subjected to a rigorous peer review prior to publication. CCPS gratefully acknowledges the thoughtful comments and suggestions of the following peer reviewers:

Robert Bartlett	Pareto Engineering & Management Consulting
Andrew Basler	Mallinckrodt Pharmaceuticals
Michael Broadribb	BakerRisk

Mark Jackson	FM Global
Morteza Jafari	ABS Group Consulting (USA)
Pamela Nelson	Solvay
Chad Patschke	Ethos Mechanical Integrity Solutions
Perianan Radhakrishnan	Petrochemical Corporation of Singapore
M.S. Rajendran	ABS Group Consulting (Singapore)
Darrell Wadden	Nova Chemicals Ltd.
Dan Wilczynski	Marathon Petroleum Company
Della Wong	Canadian Natural Resources Ltd.

PREFACE

The process safety community, through professional and industry associations, has focused considerable attention on Asset Integrity Management (AIM) of equipment directly involved in process operations. The purpose of this book is to address integrity management of assets that often fall outside the traditional process safety management asset integrity program, because they are not ranked high as "safety critical" and have long lifecycles. In particular, such assets include process supporting infrastructure like pipe racks and bridges, equipment supporting structures, sewer and drain lines, rail spurs, and process buildings to name a few. Failure of these types of assets can be contributing factors to process safety incidents and should not be ignored.

Aging process equipment, facilities and infrastructure are common in industry today. The developed world has expanded at an ever increasing rate placing high demands on our existing infrastructure. In many instances, equipment is now required to operate at conditions well beyond those anticipated in the original design. Service life may also have been extended. The option to retire and replace aging equipment is often not practical or economical. In fact, sometimes decisions are made to run equipment to failure.

Industry needs to better manage what it has built and acquired over the past several decades. There is no established set of rules for doing this. Each company or operating facility must examine its own business practices and goals and determine a strategy that meets its own risk criteria.

Aging equipment presents a challenge to managing the integrity of plants and associated infrastructure. This book examines the concept of aging equipment and infrastructure in high hazard industries. It specifically looks at the causes and effects of aging in many types of facilities. Possible options for dealing with the problem are highlighted without providing prescriptive advice. Related publications from the Center for Chemical Process Safety (CCPS) and others are cross referenced to provide the reader with a better understanding of the problems encountered by others and some of the solutions that have been applied. The challenge of dealing with aging process facilities and infrastructure is merely one component of a "broad based" Asset Integrity management program. The material herein was developed and compiled by a team of industry practitioners to supplement and expand upon the discussion of aging facilities and infrastructure in the CCPS publication "Guidelines for Asset Integrity Management".

The American Institute of Chemical Engineers (AIChE) has been closely involved with process safety and loss control issues in the chemical and allied industries for more than four decades. Through its strong ties with process designers, constructors, operators, safety professionals, and members of academia, AIChE has enhanced communications and fostered continuous improvement of the industry's high safety standards. AIChE publications and symposia have become information resources for those devoted to process safety and environmental protection.

CCPS is chartered to develop and disseminate technical information for use in the prevention of major chemical accidents. The center is supported by more than 190 Chemical Process Industries (CPI) sponsors who provide the necessary funding and professional guidance to its technical committees. The major product of CCPS activities has been a series of guidelines and concept books to assist those implementing various elements of a process safety and risk management system. This book is part of that series.

1

INTRODUCTION

All physical systems and process equipment undergo continuous change as a result of their chemical exposure, natural environment, service conditions, electromagnetic fields and gravity just to mention a few. When systems change, their physical properties and performance characteristics are often altered. Usually this alteration is one of deterioration or worsening. When there is a mismatch between assumed design properties and actual properties, system integrity may be compromised and failure may be more likely. The lifecycle of existing industrial facilities has increased over the past few decades. Many facilities are now operating beyond their intended life span and at somewhat harsher or more aggressive conditions. Consequently, aging may be more prevalent under more severe operating conditions, harsh weather extremes and an increase in the number of upsets, start-ups and outages than may have been originally planned or designed.

We generally measure age in increments of time. In fact, from a scientific perspective, time is simply a measure of change. Time can be measured in years or decades, or in the number of operating hours. Some forms of aging (e.g., metal fatigue) are measured in terms of the number of unit operating cycles a structure is subjected to. We often associate aging with deterioration. As we grow older our bodies deteriorate and we are often unable to undertake activities we enjoyed in earlier times. Physical structures and process equipment also have a tendency to deteriorate with age. In some venues aging is not necessarily viewed as negative; vintage wines often improve with age. From an aesthetic perspective society tends to value older architecture as well as ancient ruins and artifacts. However, this is not the case for Industrial facilities.

As systems age chronologically, three outcomes are possible:

1. Properties may improve
2. No change may take place
3. Properties may deteriorate

While the first two of the listed outcomes are not typical for process and infrastructure facilities and are not addressed in the book, the third represents a risk to a safe and reliable operation in the process industries.

1.1 OVERVIEW

Aging equipment presents a challenge to managing the integrity of plants and associated infrastructure. Included in this scope are chemical plants, oil refineries, power plants (including nuclear), steel mills, manufacturing plants, pipeline terminals and railways to mention just a few. Rigorous in-house methods must be employed to gauge quality and reliability at a given point in time. Second, and most important, the aging process and associated deterioration is not necessarily linear with time, making strategic decisions somewhat difficult.

As indicated, the aging process in physical systems and equipment is one associated with deteriorating properties and conditions. However, equipment aging does not necessarily correlate with chronological age or time in service. Aging does not necessarily equate to visible wear and tear, either. Given that time is not the only factor in the aging process, aging can simply be considered as negative or undesirable change that can result in diminished integrity and reliability. There are many ways in which material may react with its environment. Changes may affect the physical as well as chemical properties of the material including but not limited to the thickness, the crystalline structure, the tensile strength, the conductivity, and the ductility. Everyone associated with the operation and management of chemical facilities shares the challenge to operate facilities safely and reliably. To do so, it may require not only timely intervention to fix problems when they occur, but periodic inspections and system testing throughout the entire lifecycle of equipment change.

1.2   PURPOSE

This book is about the aging of process facilities and infrastructure. It explores some of the many ways that equipment in the process industries might age and suggests some of the warning signs for which to look. It is primarily intended to provide helpful ideas and suggestions to persons on the front line charged with the responsibility for dealing with aging equipment. The scope herein not only includes equipment in direct contact with process fluids or exposed to operating conditions but, additionally, the infrastructure that supports the operation. Included in this category are roads, buildings, support structures (pipe racks and access platforms), sewers, power lines, pipelines, tanks, silos, loading racks, marine facilities, and waste water/sewage ponds. Electrical equipment and instrumentation are also subject to physical aging as well as redundancy. This category includes conduits, cable trays, transformers and switchgear.

This book highlights a growing concern in the process industries. It is intended to enlighten the reader on some of the current issues confronting the safe operation and management of industrial facilities. It does not provide prescriptive advice for dealing with aging but suggests some ideas that might be applied as part of an Asset Management program. Many of these ideas have been tried and tested by CCPS member companies. Ultimately, some difficult decisions will need to be made to determine what equipment to replace and what equipment may continue to be operated safely. By recognizing and understanding aging it is hoped one can adopt strategies to help operate facilities in a safe and responsible manner.

Some examples of aging facilities are shown in Figures 1.1-1 and 1.1-2. While the effect of passage of time on this equipment is recognizable from external appearances, this is not always the case. Visual appearance alone is not sufficient to gage the condition of an asset.

1.3   AGING: CONCERNS, CAUSE AND CONSEQUENCES

What is it about aging equipment that should be of concern? It is the unknown and increased potential for failure, resulting in safety, environmental incidents or business interruption. Such failure may be physical or functional. Either category can have catastrophic consequences. An example of physical failure is material breakage due to pre-existing high stresses or deteriorated properties. A functional failure is one that impedes or interferes with the intended functional capacity of a system. Instrumentation and control systems are susceptible to functional failure due to aging.

Physical failures are often visible to the naked eye and there may be warning signs prior to major consequences. Sometimes physical integrity may be difficult to detect or measure through visible means. Hidden defects can contribute to a future failure. Likewise, functional failures are often less obvious and may be more difficult to detect. A physical failure can coincide with a functional failure if a system is unable to perform its required function following failure. An example of a physical failure without functional consequence might be paint peeling or deteriorating on a metal surface while the properties and performance of the metal component are not immediately affected.

Figure 1.1-1. Image of an Aging Facility Containing Silos

Figure 1.1-2. Vintage Vessels Fastened with Rivets

On the other hand, a purely functional failure might be the inability of equipment to operate at high (previously demonstrated) throughput. A system that does not perform properly when required to do so can undermine the safety and integrity of an operation. Electrical and instrumentation systems fall into this category. We depend on high availability under all situations.

Physical failure can occur in systems and equipment being exposed to forces whether they are mechanical, electrical or magnetic. The actions of chemical (both environmental and process chemicals) exposure can also contribute to physical failures. For instance, Stress Corrosion Cracking (SCC) issues on pipelines are both environmental and mechanical physical failures (stress related and time dependent).

The simplest of these forces is gravity. Gravity exerts a downward force on all equipment and upon prolonged exposure it can cause bending or sagging. If moving parts are involved, shaft alignment may become distorted causing increased wear from friction.

Structural creep is a phenomenon related to gravity whereby slight dimensional changes take place upon prolonged exposure to high loads. Structural creep is irreversible. Creep in metals occurs at a higher temperature and causes minute, incipient grain boundary melting or micro voids that cause weakening. Creep often occurs in boiler and process heater tubes (e.g., ethylene cracking furnaces). Sometimes creep is accompanied by surface or concealed cracking which may progress towards a mechanical failure. Creep may also occur at normal environmental temperatures resulting in sagging. An extended structural member or long span of piping is subject to normal gravity as well as operating loads and weather (snow and ice). Over time such components may bend or sag in response to these forces. Whether the strength of the affected components is impaired may require careful inspection and analysis.

Electric and electromagnetic forces are present in all operating equipment and structures. Electric and electromagnetic forces can alter the grain structure of steel leading to local weak spots creating the potential for failure.

Physical materials are typically chosen because of their tensile strength as well as their chemical properties. Ferrous metals are commonly employed in the process industries for vessels, piping and other equipment. Metal must be rigid and strong enough to withstand forces in an operating environment. Properties such as thermal or electrical conductivity, hardness, resistance to chemicals are some of the prerequisites to material selection. Process equipment materials may also include glass, rubber, ceramic and other metals including alloys. When the important properties of process equipment are diminished or compromised, such equipment may no longer be fit for service and a failure may be more likely to occur. Failures related to equipment aging will be dealt with in more detail in subsequent chapters.

Chemical exposure is another contributor to equipment aging. All equipment is exposed to chemical substances which constitute the operating environment. That environment may include air, steam, water, corrosive liquids and vapors, reactive chemicals, organic compounds and biomatter. Various forms of residue may accumulate in equipment and, if not properly cleaned for several years, can contribute to material degradation and plugging. This material may be difficult to remove if service conditions have changed and it has become integrated with the base metal.

The most common category of chemical change is corrosion. When two or more incompatible materials come into contact, chemical change is inevitable. Chemical change affects not only the exposed surface of equipment but it may penetrate far into the material thickness compromising the physical properties for which the material was selected. System incompatibility may result due to hybrid systems comprised of old and new components. Some incompatibility may exist contributing to confusion and human error, adjustment of dimensional discrepancies using improper materials, galvanic corrosion and electromagnetic currents (e.g., at underground and above ground piping transitions).

Many of these mechanisms may be involved in service aging. Service aging is the product of the operating history of the equipment, including failures, breakdowns and process upsets and how these operating conditions have impacted the remaining life of the equipment. This is particularly true when the equipment was originally designed based on known/expected damage mechanisms and failure modes as determined by a process hazard analysis, but during its operating life those original design conditions have changed, resulting in increased deterioration and the potential for breakdown.

Natural events can also affect equipment and structural aging. As equipment gets older it has a higher probability of being exposed to "rare" events which may not have been thought of in the design. Flooding allows ingress of water and moisture into structures with deleterious impact, or subjects vessel supports to upwards stress due to buoyancy. Drought can cause changes in the water table and impartment of equipment ground systems. Natural settling causes foundation sinking or distortion that weakens support structures. Extreme wind can exceed wind loads or cause surface erosion from blowing sand and dust. Soil creep, expansive soils, and seismic activity may produce long term cumulative effects.

Lack of documentation and knowledge is also a concern associated with aging equipment. Record keeping on older facilities was not always to the level of detail as required by current safety regulations. A written history of service conditions or upsets during the entire lifecycle often does not exist, and tenured staff with this information may have left or retired. Often older facilities have had several owners, and equipment records have been lost or misplaced. Third party leased equipment and facilities may also present these problems: unknown design basis and materials of construction, unknown history, unknown previous service, liabilities, etc.

1.4   HOW AGING OCCURS

Aging is about change and it is driven by exposure conditions and forces. Aging may involve changes in physical dimensions and appearance or it may involve changes in properties. Even if those conditions or exposures are constant, there is no guarantee that aging will progress in a linear fashion. Visual changes such as discoloration, cracking, peeling paint and other surface blemishes are often cosmetic in nature and are unlikely to pose a risk of failure. They are still indicative of change however and should trigger a more in-depth look at properties to determine if these have been altered.

Aging is commonly associated with the progression of time within the lifecycle of a system or facility. From a reliability and integrity perspective, however, aging also recognizes that physical and chemical properties may deteriorate upon continuous or intermittent exposure to normal or upset operating conditions. The fact is, even exposure to a stable or dormant environment can bring about deterioration in some systems. Such deterioration, if not detected and addressed, can make systems more prone to failure. The goal is to aim for better understanding of the process of aging so that better informed decisions can be made, and catastrophic events avoided. Figure 1.4-1 depicts an aging scale which illustrates how aging evolves from minor cosmetic defects to total destruction.

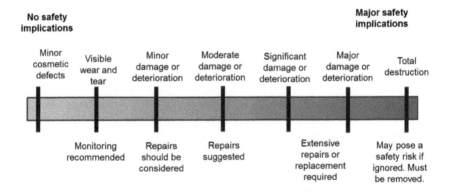

Figure 1.4-1. Suggested Spectrum for Aging Facilities

1.4.1 Metallic Corrosion

Metallic corrosion and surface deterioration are likely the most visible symptoms of aging and are familiar to all of us. Most metals including alloys react with their environment. The rate at which this occurs is a function of the base metal properties, other materials or contaminants at the point of exposure and conditions such as temperature. For simple rust to occur, air and moisture must be present. If the surface contaminants are in the low pH range, corrosion will occur at a more rapid rate. A layer of rust, if left undisturbed, can actually provide a protective layer to prevent further corrosion of some ferrous metals. Operating parameters such as fluid velocity can influence the removal of corrosion products thereby exposing base metal and further aggravating the problem. What's interesting about corrosion is that, once it begins it is difficult to stop it even if the exposure is controlled. Removal of surface deposits down to base metal is often required but this causes harm since it reduces the metal thickness. Unless surface protection such as paint is applied and efforts are made to control the exposure, further deterioration can occur. It is also important to mention that episodic corrosion due to exposure from an accidental release of a corrosive material (such as acid) can also take place, and may damage the surface of the infrastructure. Having proper spill prevention programs in place may prevent infrastructure deterioration caused by releases of corrosive. Also, corrosion products which are toxic or pyrophoric may be classified as hazardous waste and may require regulated disposal procedures.

While rust is a common form of corrosion found on the surface of ferrous metals, there are many other related corrosion mechanisms that can also hinder the properties of various metals. These include pitting, fretting, stress-corrosion cracking, galvanic corrosion, hydrogen attack and sulfidation. Each mechanism has its own symptoms and causes, and a thorough knowledge of metallurgy is required to ensure that systems are designed to match all anticipated operating conditions. In the past, some facilities may have been designed and built without the benefit of such knowledge. Furthermore, as operating conditions evolved over later years, a mismatch between design and operation may have occurred without being recognized.

The effects of metal corrosion are more than just a reduction in material thickness. Corrosion can alter the physical properties of metal. Ductility, hardness, porosity and electrical conductivity are just a few of these. A corroded or tarnished surface of an electrical contact point can act as an insulator and may contribute to a power interruption. Electrical fixtures and instrument boxes contain thousands of metal contacts which are subject to aging and deterioration. These can suddenly fail to conduct current without prior warning. They must be opened and inspected at regular intervals.

There are many different types of corrosion and the causal mechanisms vary widely. It is difficult, if not impossible, to find a metal or alloy that is not vulnerable to some type of corrosion. Prudent material selection during the design phase provides the best opportunity to combat the problem. It is beyond the scope of this book to examine all the various types of corrosion and their effects on metal properties. However, one fact is noteworthy. Corrosion seldom occurs at one local point in a system. When corrosion is directly encountered or corrosion products are discovered in a process stream, every effort should be taken to thoroughly inspect all parts of the system to determine the cause and the extent of the problem. Better still, conduct these inspections early in the lifecycle of new systems and continue this activity on a planned basis.

It may also be worthwhile to consider the opposite issue, that is, if corrosion is expected but is not observed as expected. It is important to continue to investigate why the expected corrosion is not appearing. It could be that the corrosion, or a related corrosion mechanism, is appearing at a different location.

1.4.2   Corrosion Under Deposits

Some piping systems are more prone to accelerated aging due to corrosion. Leaks can occur in older underground pipelines due to lack of cathodic protection and deteriorating coatings resulting in release of contaminated water, hydrocarbons and sewage. Corrosion Under Deposits (CUD) occurs in lines that are partially plugged with deposits such as slime, sludge and sediment as contaminants concentrate under the solid deposits and cause more rapid corrosion. This is a particular concern in upstream petroleum produced fluid gathering system piping.

1.4.3   Corrosion Under Insulation and Fireproofing

Corrosion Under Insulation (CUI) is a possibility when the equipment temperature is cool enough to condense moisture. Failure of the outer insulating cover and coating, as well as ingress of rain water can also cause of CUI. CUI is an aging issue for refrigeration equipment and insulated piping, especially in coastal regions with high humidity and salty air and the equipment temperature is cool enough to condense moisture and not cause frost. The insulation becomes water logged and keeps the moisture in contact with the metal, causing more rapid corrosion which is exacerbated by chlorides in the atmosphere or in certain types of insulation. CUI can potentially cause stress corrosion cracking in equipment and piping constructed of austenitic and duplex stainless steels due to chloride attack.

Deterioration of insulation coverings, coatings and fireproofing can allow the ingress of water (rain, cooling tower mist, deluge testing discharge) that leads to CUI. CUI is of particular concern because it is not easily detected with typical inspection techniques. In some instances, it is only detected after considerable damage has occurred.

CUI is not only an aging mechanism that occurs at low temperatures. Hot pipe and equipment are sometimes insulated, not only to conserve energy, but also to protect personnel. Another reason for CUI is the use of steam tracing. When steam tracing is used within insulation, extra precautions are needed to resist stress-cracking corrosion. Steam traced systems experience tracing leaks, especially at tubing fittings beneath the insulation. Figure 1.4-2 shows an image of external corrosion of a pipe due to leakage of steam tracing.

Figure 1.4-2. External Corrosion of a Pipe Due to Leakage of Steam Tracing (Sastry, 2015)

Steam-traced lines should be double wrapped, with the first layer applied directly to the pipe, followed by the steam tracing and then more foil over the top.

Corrosion under fireproofing (CUF) is caused by moisture being trapped under fireproofing used on structural steel supporting process equipment including pressure vessel skirts, structural platforms, pipe rack structural steel and sphere legs. Cracking or spalling of fireproofing over time allows ingress of moisture. Like CUI, CUF can be more dangerous because it often goes undetected. It is difficult to inspect and the consequences of failure may be greater.

For more discussion of CUI and CUF mechanisms the reader is directed to API Recommended Practice 583: *Corrosion Under Insulation and Fireproofing* (API 2014).

1.4.4   Manufacturing Defects

Manufacturing defects are another potential contributor to aging. A casting flaw can remain hidden for years and may suddenly trigger a fault line leading to mechanical failure in process equipment e.g., pump casings, valve bodies). With large size equipment, such defects are becoming more common. At least one significant incident in a large power plant occurred when a generator set disintegrated at high speed revealing an irregular crystal inclusion in a large drive shaft. This defect had remained hidden for years and a slight vibration problem was masked by counter balancing during initial commissioning. It may be important to fully investigate any suspicious properties or behaviors in equipment during initial commissioning since there may not be a second chance.

1.4.5   Excessive Wear and Tear

Wear and tear is another significant contributor to aging of equipment and infrastructure. Wear and tear is damage that naturally and inevitably occurs as a result of normal exposure to service conditions, upset conditions and/or extended time. Wear and tear is not necessarily a function of chronological age but rather the sum total of exposure conditions through the entire lifecycle. Despite efforts taken during design to anticipate all operating conditions, there are often abnormal (out of range of the design envelope) situations that may not be anticipated. Sometimes these are recognized while at other times they are simply missed or overlooked. These abnormal conditions can have a pronounced effect on aging.

Wear and tear is an expected phenomenon in most mechanical equipment. When equipment is directly exposed to operating and environmental conditions it undergoes change. This change may be detrimental to material properties and can lead to premature failure. The design lifecycle of mechanical equipment should consider all the factors that might contribute to wear and tear during normal and occasional upset conditions. Experience with similar systems can provide valuable insight into making such an assessment.

There are situations where excessive wear and tear is experienced. This may be the result of substandard materials of construction, improper assembly or operating exposures that were not anticipated. Excessive wear and tear should be seen as a red flag that something is out of the ordinary. Unless the mechanism contributing to such deterioration is well understood, an unexpected failure could occur at any time. Linear extrapolation of remaining life should only be applied when wear and tear correlates with past experience. Similarly, an unusual pattern of minor failures not experienced elsewhere or previously should signal that something is wrong. Increase the frequency and intensity of monitoring and consider early replacement. Better still, analyze the operation and determine the cause of the excessive wear and tear.

What is excessive wear and tear and how is it recognized? There is no simple answer that can apply to all situations. The first step is to establish what is normal based on the review of the process hazards where expected damage mechanisms are evaluated as part of the equipment design for the process. This requires a comparison against a similar system or piece of equipment in a similar operating environment. This may be difficult for unique or prototype equipment. If such a reference point is not available, one must apply experience and judgment. Another practical approach might be to consider visible markings (etching or scoring), damage, or deterioration beyond a certain percentage of that normally expected as excessive wear and tear for similar equipment regardless of service. Ultimately, the definition of excessive wear and tear and the associated response will depend on the operational risk and the willingness to tolerate a failure. Figure 1.4-3 shows scoring on a shaft.

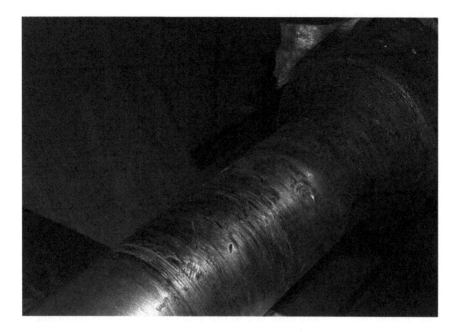

Figure 1.4-3. Image Showing Scoring on a Shaft

1.4.6 Fatigue

Fatigue is the exposure of a structural member or component to a high number of stress cycles. Cyclic stressing is probably the single biggest contributor to aging other than corrosion. Cyclic stresses due to loading, start-ups, pressure swings, mechanical impact, temperature cycles, and wind gusts can have a more pronounced effect on equipment integrity than continuous high stress. Effects may include cracking and fatigue failure.

Cyclic stress is commonly associated with vibration in machinery and equipment which operates under repeated high load conditions. Fatigue itself is not necessarily a problem and is an expected behavior in certain operating systems. However, if not monitored, it can result in an abrupt fatigue failure. Metal fatigue is typically characterized by a breakage pattern in a single flat plane similar to a shear. Fatigue is one of the prime reasons that commercial airlines retire their aircraft after a specified number of flights. Incident statistics have shown that defects are more likely to occur in short haul aircraft exposed to more frequent take-off and landing schedules. In process facilities, cyclic stresses may be due to start-ups, pressure swings, and upset conditions.

1.4.7 Non-Metallic Aging

Material aging extends beyond corrosion and applies to many different materials. Rubber and plastic may gradually dry out and lose their properties. Surface cracks may develop and extend through the entire thickness of the material. The problem is further exacerbated when rubber is required to bend or stretch repeatedly. A rubber drive belt can dry out. Belt slippage may contribute to overheating and eventual failure. Hoses exposed to high pressure fluids or prolonged exposure to ultraviolet light can disintegrate and fail without prior notice. Even some ceramic materials are prone to deterioration as a result of prolonged exposure to other substances. Insulation on electric wiring can harden and break off. Insulation failures are common after several years of service and these can cause an electric short or ground fault. Wood and other non-metallic construction materials can rot, chemically decay or be attacked by insects again destroying the properties for which they were chosen. Wood, in particular, is subject to warping and twisting as it dries.

Glass fiber re-enforced piping is commonly used in cooling towers and fire water systems. It is very strong and robust particularly when it is loaded in a direction parallel to the fibers in the material. However, internal stresses can occur during the curing process resulting in distortion and cracking several months after the resin has set. This is particularly true with complicated shapes or manifold configurations. Studies have shown that glass fiber materials can lose 60% of their strength after one to two years of service. It is important to inspect glass fiber re-enforced piping for cracks and leaks on a regular basis. If such systems are in highly hazardous service, consider replacement with an alternate material.

Concrete is another important material commonly found in our industrial infrastructure. Although concrete seldom comes into direct contact with chemicals, electrical energy or other high energy sources, concrete is still subject to aging. A structural failure can be catastrophic. Concrete is somewhat unique in that its compressive strength actually increases with age during its initial curing cycle. This assumes that it is properly mixed. Under full and variable load conditions concrete can spall and erode. Ultimately, it will deteriorate from the outer surface and as well as internally. Exposure to road salt and other chemicals can accelerate the aging process in concrete. If the internal rebar is exposed to a corrosive environment, it also can decompose and contribute to a failure.

Passive fireproofing other than concrete is often used to protect structural members from the heat effects of prolonged pool fires. Cementaceous fireproofing has a finite life and it can crack on exposure to climatic conditions as well as mechanical impact. In a fire situation, it may not stand up to a water jet from a fire hose or monitor. Proper installation, inspection and repair should be practiced avoiding premature aging of structural fireproofing.

Common construction materials such as widow glazing and joint calking may also pose aging issues. Some window glazes contain asbestos, which may become airborne dust when the glazing dries out with age. Polychlorinated Biphenyls (PCBs) may exist in caulking materials which can leach into building material (cement and bricks), necessitating expensive remediation and disposal.

1.4.8 Aging of Physical Structures

Physical structures such as buildings, offices and shelters age as a direct result of their exposure to the natural environment. Wind speed and direction, weight of snow, hurricanes, tornadoes, earthquakes, temperature variations and precipitation can cause structural members to move or separate and building materials to deteriorate. Foundations can shift or sink causing superstructures to collapse. Doors and walls may fall out of alignment. Changes in the water table and soil erosion can compound this problem. Leakage through roofing materials can lead to water damage and mold. Ultimately, a facility might need to be abandoned for health reasons. Timely inspections and upkeep can prevent this from happening. When building repairs are deemed necessary they should be scheduled and completed. Otherwise, deterioration will continue and repairs may not be possible or practical.

Aged construction materials used in buildings themselves could represent a hazard. Are there any walls that contain asbestos or are constructed with asbestos panels? If so, these need to be removed to avoid personnel exposure. Careless removal of asbestos by untrained workers can create a greater hazard than if the asbestos were left intact. Ensure asbestos is removed properly and that all safety regulations are followed.

On the electrical side, are there any transformers or capacitors that still contain Polychlorinated Biphenyls? These fluids were previously used as a coolant and dielectric medium but are highly toxic. This material is also classified as a carcinogen. PCBs were eventually outlawed in the power industry and a substitute material was implemented. However, many PCB contaminated sites still exist in the United States of America (USA). Another hazardous material is lead based paint. The scraping and removal of this can release dust particles into the air presenting a hazard to nearby workers. Investigate whether less hazardous coatings have been applied to all plant equipment.

Roads, ditches, power lines, energy pipelines, cooling towers, operating plants, laydown yards, waste ponds, rail lines, pipe racks, conduit racks and marine facilities constitute important infrastructure for industry. A failure in any of these systems can prove costly and can directly impact the community. Of some concern is the fact that more money is spent on new (additional and add-on) facilities than on the existing infrastructure. Infrastructure is often not viewed by corporations as a direct contributor to profits. The common contributor to infrastructure aging is weather and ground movement. Rain, snow, ice and wind place thermal and mechanical stresses on the infrastructure systems listed above. Are any power lines supported on wooden poles and, if so, are they securely anchored? Has the wood been inspected and is it free from rot? Roads and ditches in particular are subject to erosion and flooding. A damaged road surface will contribute to vehicle damage and repeated usage will aggravate the problem. Proper repairs may be necessary to address current service conditions. Underground services are often exposed to unknown conditions. In many regions, the water table has shifted over time causing soil movement to occur. Many construction excavations have met with surprise to find damaged or shifted equipment below.

1.4.9 Process Chemicals Aging

Chemicals can also deteriorate with time. Slurries and chemical mixtures break down and decompose upon exposure to air and changing exposure conditions. Phase change may also occur in which a gas is released or two liquid phases separate out. Obviously, the chemical properties will also change when this occurs. Lubricants, transformer oil and heat transfer fluids are examples of chemical substances that support an industrial operation but do not come into direct contact with process fluids or product streams. As such they are an important part of the infrastructure. Sometimes these materials are corrosive or they may react directly with certain metals. If these chemicals are overlooked or ignored, they can harm the equipment in which they are contained or enclosed. Even more, if a spill takes place, accelerated corrosion can occur if the surface is not properly cleaned. It is important to regularly monitor the quality of all chemical substances whether or not they are part of the main process. If contamination or quality deterioration are noted, these materials should be removed and replaced. If equipment containing chemical solids or fluids is to be out of service for an extended period, consider replacing the contents with a more inert material. Again, regular monitoring is essential. Biohazards have recently been recognized as a significant threat to the health and safety of workers (dry rot, mold, asbestos etc.).

1.4.10 Aging of Specialized Equipment

Pumps, compressors, turbines and other specialized mechanical equipment are often comprised of several moving parts. Over the course of a normal lifecycle many of these parts will have been replaced. Seals, impellers, bearings, gaskets and wear plates likely fall into this category. However, significant questions remain. Even with good equipment records, is there assurance that the entire unit has been changed out and replaced? Are there any original parts or components that have not been inspected and potentially could fail at some point in the future? Could a pump casing or a turbine rotor fail as a result of several years of operation? Inspection programs need to be extended into those hard to reach parts of an operation to ensure that equipment is safe, reliable and continues to meet operating design parameters. A Management of Change system is needed to manage these changes and to verify compatibility with the equipment design for the intended service.

1.4.11 Obsolescence

Obsolescence is a major concern and important consideration with aging. Obsolescence suggests that a system, facility or piece of equipment no longer meets current requirements or is technically incompatible with its surroundings. Obsolescence is often the result of technical advancement and innovation. Why is this of any significance? If two or more parts of a system do not function harmoniously an incident may be more likely to occur.

Obsolescence may also be the result of inability to obtain spare or replacement parts. For example, the advancement of electronic controls over pneumatics, use of fiber optics instead of low voltage electrical, and computer consoles with graphic displays, instead of board mounted instrument panels. As technology develops, vendors and suppliers typically promote their latest products and ultimately abandon earlier designs. As customers fall into line with newer technology, there is a reduced demand for "older" equipment and associated spare parts. Simple economics dictates that this will signal the end of the line. To combat the problem of obsolescence, industrial operations will often substitute "new improved" parts or they may attempt to manufacture replacement parts in-house.

Another strategy that is often employed is to scavenge spare parts from used equipment vendors, or cannibalizing from redundant in-house equipment. This was sometimes practiced by foreign domestic airlines flying aging aircraft and resulted in above average carrier accident history statistics. The challenge is that the service history of these parts is generally unknown and even with systematic inspection, conditions like metal fatigue present unknown or unacceptable risk without a history of the number of stress cycles. Also, counterfeit or used parts may not meet specifications and tend to only compound the problem. By allowing such practices, a company may be knowingly or unknowingly accepting a high level of risk.

There was a case involving a company division, whose product line profitability was in decline. The Waste Water Treatment (WWT) facility at one of the division's plants was antiquated with pneumatic control systems that were no longer supported by the manufacturer. A corporate integrity review team had flagged the issues, but the system was given a low priority for replacement. The maintenance department had to resort to cannibalizing and "replicating" spare parts to keep the facility running. Due to the makeshift spare parts degrading the functional integrity of the controls, there were frequent effluent concentration excursions.

But "new improved' parts are not necessarily the solution to obsolescence. These are often designed to match new equipment that may have different performance characteristics. New parts must also meet current industry standards which again may be somewhat different from those in place during original equipment design. New improved parts must be compatible in a number of key areas including size, units of measure (English or Metric), materials of construction, joints and fasteners, functionality, software and sensitivity to environmental conditions.

There are some simple concepts to remember. Substitution introduces something that is different. Difference equates to change. A significant number of incidents in the process industries trace their roots back to change and an ineffective "Management of Change" system. To conclude this discussion, it is wise to reflect on the simple fact that: "newer and better" is not always safer!

1.4.12 Redundancy

There are different definitions for the concept "redundancy". The first concept refers to a term closely related to obsolescence. In this case, redundancy suggests that equipment is no longer required. This is often not related to the condition of the equipment but to the market or business needs. Such equipment or facilities may then be retired from useful service or left in a standby condition to support a system failure. In a standby mode, it is essentially in service and it is typically configured into an operating system by means of piping or electrical power. This equipment should be available for full service on short notice. An example of retirement mode might be a registered pressure vessel that is deemed as surplus equipment. Obsolete equipment often falls into the redundant category but the converse is not necessarily true.

Equipment becomes redundant when it no longer meets the business or service requirements of an operation. Common causes of equipment redundancy include inadequate size or capacity as well as inability to meet current operating conditions (temperature and pressure). If redundant equipment cannot serve a useful purpose in the future, it should be dismantled and disposed of in a responsible manner. One precaution when removing redundant equipment is that "dead legs" created in the remaining isolated piping need to be eliminated. Such dead legs are susceptible to failure due to freezing, vibration and accelerated corrosion.

Simple abandonment is not acceptable in many jurisdictions. Abandoned equipment is a magnet for vandals and wildlife. It can also breed biohazards which are harmful to workers and the general public. Abandoning equipment in place also carries with it some potential negative side effects such as: 1) Potential damage to operating equipment if there is a structural failure of abandoned equipment. 2) Overall impact on psyche of the workforce because of the physical appearance of the abandoned equipment. Many operations store redundant equipment in a laydown yard for possible re-use in a future application. If re-use fails to occur, redundant equipment may capture some value as scrap through a waste recycler.

The second meaning of the word "redundancy" refers to a part in a machine or equipment that has the same function as another part, which leads to greater reliability and having the spare or stand by part available. In this case, the equipment may be considered redundant but not yet obsolete. If redundant equipment might serve a secondary backup purpose or may enter into a new lifecycle it must be carefully managed and included in an Asset/Mechanical Integrity Management program. Left idle for a considerable period of time it will be continually exposed to changing environmental conditions and could further deteriorate. In fact, sometimes this deterioration may be more serious than if the equipment had been left in service. This may make it unsuitable for future use. Redundant equipment should be cleaned and properly sealed off to prevent unauthorized tampering or entry. Consideration should be given to introducing an inert purge or medium. The equipment should also be inspected, tested and maintained as part of the asset integrity program at regular intervals to gauge whether deterioration is taking place to verify the equipment will be fit for service if needed.

Equipment components and service items that have never been used are also susceptible to aging. This includes spare parts, stored equipment and chemicals and fluids in warehouses. Long, large, replacement rotors for compressors and turbines may deform. Electronic components may degrade due to environmental conditions. Lubricants have a shelf life and will deteriorate with age, hindering their properties.

1.4.13 Brownfield Construction

The challenges of brownfield construction are all around us. Brownfield construction involves physical work in an area currently or previously occupied by process facilities and equipment. Sometimes the work is done while existing equipment is fully operational. Seldom are new operating facilities constructed in an untested region. Regulatory approvals, cost and lack of skilled manpower have hindered greenfield construction. The lure of "adding to what currently exists" has become the current trend in industry. While this strategy is somewhat questionable from a risk consolidation point of view, it sometimes helps to achieve a level of consistency in the way facilities are operated. The benefits seem to end there. Building new on top of old extends the life expectancy of existing facilities. Apart from compromises in spacing guidelines, brownfield construction often requires numerous physical tie-ins to live equipment and existing support systems. Is the existing equipment and infrastructure suitable to support a major capital project? Are flare lines and sewers adequately sized to handle an upset condition or release? Can new piping be welded into existing piping circuits? Is there enough metal thickness remaining to ensure proper penetration welds? Will electrical systems and instrumentation be compatible with those in the existing facility? Is there enough strength and capacity in the pipe racks and cable trays? More important, has a thorough condition assessment been conducted on all equipment that will possibly need to support altered operating conditions for an extended time period?

Aging equipment, facilities and infrastructure will always be a reality in our changing society. We all have a responsibility to pitch in and contribute to making it safe and reliable. We must avoid the urge to focus only on the newly constructed or newly acquired facilities which often have more desirable features. Unless there is a compelling safety or economic driver, the lifecycle of aging equipment and infrastructure can and should be considered for future extension. We as individuals, our companies, industry and regulators should report any equipment problems or deficiencies to those in authority so that timely maintenance and repairs can be provided.

2

AGING EQUIPMENT FAILURES, CAUSES AND CONSEQUENCES

This chapter illustrates how equipment aging, if not dealt with early can translate into major loss (e.g., public and employee safety, economic loss, regulatory repercussions, environmental consequence, damage to company reputation and impact to our customers).

Aging is not about how old plant and equipment is. It is about its condition, the service it is in and how that is changing over time. The issue of aging industrial assets is of increasing importance to regulators and the industry as a whole, as its successful management is critical to the overall safety performance of process plants (IChemE, 2013).

2.1 AGING EQUIPMENT FAILURE AND MECHANISMS

There are two broad categories of failure, physical and functional. Physical (mechanical failure) is often associated with breakage and is the result of internal or external forces. Physical forces, overpressure and over- or under- temperature resulting from relief scenarios, may cause movement or they may generate stresses in equipment. These stresses may be more than the load bearing capacity of materials, making them more prone to failure. Ultimately, something has to give way. Cyclic stresses resulting from vibration, pressure/vacuum service or intermittent service (e.g., refinery coke drums) can cause fatigue cracking that ultimately penetrates through the thickness of the metal. Metal fatigue often results in sudden or abrupt mechanical failure. Physical failure is often associated with movement or deformation such as bending or stretching. Sometimes the material may separate or fracture. There may be one slight advantage associated with this type of failure, in which it might be possible to predict the failure and prevent it based on physical symptoms.

Functional failure is somewhat less tangible than physical failure and it is often more difficult to predict. Functional failure is the inability to perform a service or function when required to do so. Power outages, loss of containment (e.g., pipeline leaks, bolted connection leaks, electrical flashovers) virus infection of computers and instrument malfunctions are examples of functional failure. Again, these may appear suddenly without warning. Near misses and chronic problems are a possible forewarning of system failure and can be used as both leading and lagging indicators (CCPS, 2009). Functional failure may also be the result of dimensional problems and changes in properties. An engine will not operate smoothly when the rotating elements are worn and ultimately it will fail in service.

Which category of failure is the most significant in terms of potential consequences? There is no single answer. At the macro level both categories of failure can have disastrous consequences. A falling or collapsing structure resulting from a failed structural member can cause serious injuries and fatalities. A pipe or vessel failure in a chemical plant can release hazardous material as well as energy, again with tragic consequences, "including the potential for a fire and/or explosion following the release of the stored energy. The role of the asset integrity program is to keep this energy stored where it belongs. On the functional side, we place high dependency on the continuous availability of systems such as electrical power. A surprise failure can compromise or interrupt critical activities, again causing significant physical damage and business interruption. A case in point, the failure of a safety critical system allows a less serious event to become catastrophic within a high hazard facility. Physical or mechanical failures, are likely the failure modes that have most people concerned. When the limits of elasticity for any material are exceeded the material is more likely to fail. Mechanical failure is accompanied by an exchange of energy. In large systems and equipment, the level of energy released during failure may have catastrophic consequences.

It is not only externally applied forces that contribute to mechanical failures. The natural force of gravity acts on all structures and mechanical systems. A bending moment is exerted by the dead weight of each mechanical component. Even if a piece of machinery or structure sits idle for decades, elongated components may become distorted due to gravity. Such is the case with long shaft rotating equipment where the drive shaft is known to have sagged under its own weight. To counter this problem, many operations ensure that all equipment is regularly used and is inspected tested and maintained as part of the asset integrity program. Vintage vehicles stored on display for long periods are seldom drive worthy without extensive repairs and modifications.

2.2   CONSEQUENCES OF AGING EQUIPMENT INCIDENTS

To help inform and prioritize their approach to the topic of aging plant, the Health and Safety Executive (HSE) conducted an extensive analysis and review of various United Kingdom (UK) and European-wide accident and incident data to assess the extent to which aging mechanisms are contributing to accidents and losses in onshore chemical plants over the period from 1980 to 2006. One of the databases accessed for the study was the European Union Major Accident Reporting System (MARS) operated by the European Commission Joint Research Centre.   Fatality, injury and loss statistics extracted from that data are summarized in Tables 2.2-1 and 2.2-2.

A review of the accident data reveals on average there were 1.6 fatalities and 6.1 injuries per aging equipment plant incident. Furthermore, the average direct financial loss per aging equipment incident was €1.8 million, excluding ecological and community impacts. Considering the loss in terms for human and financial impact indicates there is much room for improvement.

This study also provided insight into the causes of failure incidents based on Technical Integrity, Electrical, Control and Instrumentation (EC&I and Human Factor/Procedural type issues shown in Figure 2.2-1. The breakdown of the technical integrity category is shown in Figure 2.2-2 (HSE, 2010). Physical failures are dominated by corrosion/erosion mechanisms, and the next largest category is EC&I related failures. Using the aging only data in Table 2.2-2, the breakdown between physical integrity aging and EC&I aging is 59% and 22% respectively. While the causal breakdown for physical integrity is not available from Table 2.2-1, one could expect a similar pattern as the total data set.

Table 2.2-1. Deaths and Injuries Statistics for MARS Reportable Major Accident Hazard Incidents

Class	Total	Deaths			Injuries		
	No. Incidents	Incidents	Total Deaths	Deaths per event	Incidents	Total Injuries	Injuries per Event
All events	348	57	124	2.2	140	4,201[1]	30.0
(excluding[1])					139	1,959	14.1
All integrity	149	11	35	3.2	51	768	15.1
Integrity aging[2]	57	3	4	1.3	21	125	6.0
C&I aging[3]	21	2	4	2.0	4	32	8.0
Other aging[4]	23	2	3	1.5	7	47	6.7
All aging[5]	96	7	11	1.6	30	183	6.1

Notes:

[1] One incident with 2242 injuries.

[2] Aging from corrosion, erosion, stress corrosion cracking, fatigue, corrosion fatigue, vibration and wear.

[3] Aging where Control and Instrumentation (C&I) is a factor.

[4] Aging from other sources (safeguards, structural, etc.).

[5] All aging sources (this total is five less than the sum of the three above categories as five of those incidents are 'double counts' since classified in two or more aging categories from notes 2,4).

Table 2.2-2. Total Losses (Million € Equivalent) for MARS Reportable Major Accident Hazard Incidents

Class	Total Losses (M €)	Incidents Where Loss Occurred (no.)	Average Loss per Event (M €)	All Incidents (no.)	Average Loss Per Event (M €)
All Events	794.7	107	7.4	348	2.3
All Integrity	329.1	42	7.8	149	2.2
Integrity Aging	149.7	18	8.3	57	2.6
C&I Aging	17.8	7	2.5	21	0.8
Other Aging	3.6	4	0.9	23	0.2
All Aging	171.0	28	6.1	96	1.8

Source: European Union Major Accident Reporting System (MARS) operated by the European Commission Joint Research Centre

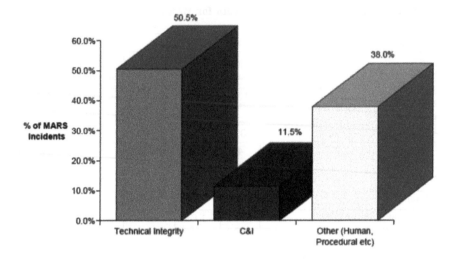

Figure 2.2-1. High Level Categorization of MARS Incidents

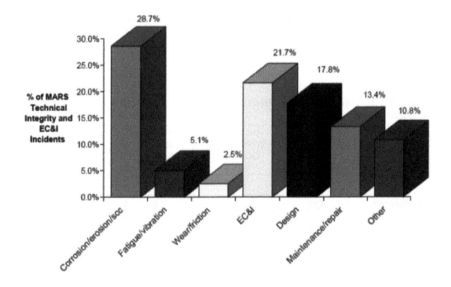

Figure 2.2-2. Causes of Technical Integrity Incidents in MARS Data

2.3   MECHANICAL FAILURE OF METAL

2.3.1   Deformation of Materials

Mechanical deformation is a stress-strain relationship depicted in Figure 2.3-1. As stress is applied to a solid ductile material, the object typically deforms (strains) in a linear fashion up to a point and in this region the strain is referred to as elastic deformation. The modulus of elasticity or Young's Modulus is the slope of the elastic deformation stress-strain line. During linear deformation, an object will return to its original shape when the stress is removed. As the stress is increased beyond a certain point, the strain transitions into a region where the deformed state is more permanent and is referred to as a plastic deformation. The stress level at the transition point between elastic and plastic deformation defines the Yield Strength of a material. In the plastic deformation region, strain hardening also occurs, caused by the accumulation of energy in the material. During strain hardening the material becomes stronger through the movement of atomic dislocations, and the deformation is no longer linear with increasing stress. Plastic deformation is irreversible, and the object does not return to its original shape. Eventually, the stress reaches a maximum which defines the Ultimate Strength. At this point the material begins to neck down (reduction in cross section area) and ultimately fails.

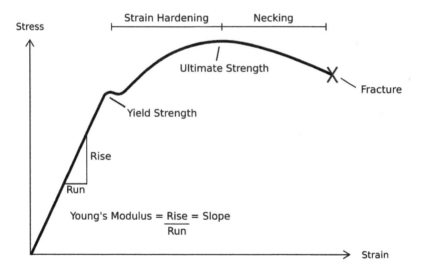

Figure 2.3-1. Typical Stress vs. Strain Diagram Indicating the Various Stages of Deformation

Source : https://en.wikipedia.org/wiki/Deformation

2.3.2   Ductile vs. Brittle Fracture

Equipment and structures fail when fracture micro mechanisms are initiated in the case when increasing forces are applied to engineering materials and components. According to the amount of plastic deformation involved in these processes, the fracture events can be categorized as ductile, quasi-brittle or brittle. (Pokluda, 2010).

Ductile Fracture. Most metals that are not too cold will exhibit ductile fracture after extensive plastic deformation ahead of crack formation. The crack is considered "stable" and resists further extension unless the applied stress is increased.

Brittle Fracture. Ceramics and cold metals are subject to brittle fracture, which occurs with relatively little plastic deformation ahead of cracking. The crack is "unstable", and propagates rapidly without increase in applied stress. A crack often propagates by cleavage breaking of atomic bonds along specific crystallographic planes (cleavage planes).

As temperature decreases, a ductile material can become brittle. The point at which this occurs is the ductile transition to brittle temperature (DTBT). For metals, the composition has an effect on the DTBT and alloying usually increases the ductile-to-brittle transition. Metal hardness has a similar effect. Therefore, brittle fracture can occur at higher temperatures than those typically associated with mild carbon steel.

Quasi-Brittle Fracture. Many quasi-brittle fractures occur as a consequence of pre-existing corrosion dimples, large inclusions or fatigue cracks. However, the localized plastic deformation at favorable sites in the bulk also enables the creation of microcracks as nucleators of quasi-brittle fracture in solids which do not contain any preliminary defects. At phase or grain boundaries, it can be accomplished by many different and well known micro mechanisms conditioned by the existence of high stress concentrations in front of dislocation pile-ups (Pokluda, 2010).

2.3.3   Metal Fatigue

Another deformation mechanism is metal fatigue, which occurs primarily in ductile metals. It was originally thought that a material deformed only within the elastic range returned completely to its original state once the forces were removed. However, faults are introduced at the molecular level with each deformation. After many deformations, cracks will begin to appear, followed soon after by a fracture, with no apparent plastic deformation in between. Depending on the material, shape, and how close to the elastic limit it is deformed, failure may require thousands, millions, billions, or trillions of deformations. Metal fatigue was a major cause of aircraft failure before the mechanism was well understood. It is still a phenomenon that is closely monitored in the airline industry today.

In materials science, fatigue is the weakening of a material caused by repeatedly applied loads. It is the progressive and localized structural damage that occurs when a material is subjected to cyclic loading. The nominal maximum stress values that cause such damage may be much less than the strength of the material typically quoted as the yield stress limit or the ultimate tensile stress limit. Fatigue occurs when a material is subjected to repeated loading and unloading. If the loads are above a certain threshold, microscopic cracks will begin to form at the stress concentrators such as the surface, Persistent Slip Bands (PSBs), and grain interfaces. Eventually a crack will reach a critical size, the crack will propagate suddenly, and the structure will fracture. Metal fatigue is also cumulative. Therefore, equipment that has passed non-destructive examination, such as a magnetic particle inspection, only proves that no microcracks have occurred yet. It provides no insight on how many more cycles the equipment can withstand before cracks appear.

Metal fatigue presents a particular problem when the presence of microcracks is detected. Since metal fatigue failure is strongly correlated with the number of stress load cycles to which a member is ultimately exposed, unless there is a recorded history of the cyclic loads, the remaining useful life is difficult to predict. Sensors can be installed to record load cycles over time. Vibration sensors and analysis to obtain a stress-strain history on large rotating equipment is often employed in the process industry.

Equipment that is exposed to cyclical stress or vibration that produces repeated elastic deformations over time can be susceptible to metal fatigue. This can lead to pre-mature deterioration that is not related to the actual age of the equipment or structure.

2.3.4 Corrosion/Erosion

Corrosion does not stand for a single phenomenon but is a generalized term to cover a destructive attack on a metal as a result of either a chemical or electrochemical reaction between the metal and various elements present in the environment. For instance, iron is converted into various oxides or hydroxides when reacting with the oxygen present in air/water, when in contact with a more noble metal such as tin, or when exposed to certain bacteria (EC JRC, 2013). An international standard defines corrosion more specifically as a "physicochemical interaction between a metal and its environment which results in changes of the properties of the metal and which may often lead to impairment of the function of the metal, the environment, or the technical system of which these form a part." (ISO 8044, 1999). Erosion can be defined as an accelerated form of corrosion driven by shear forces of fluid moving through a vessel or piping system at higher than normal velocity. The fluid movement removes products of corrosion exposing bare metal to further attack.

There are a variety of corrosion mechanisms some of the more usual forms are briefly summarized in Table 2.3-1.

The purpose of this book is not to provide an in-depth treatise on corrosion/erosion, but rather to broadly provide context for the discussions that occur elsewhere in the book. API RP 571: *Damage Mechanisms Affecting Fixed Equipment in the Refining Industry* (API 2011) is a resource for different types of damage mechanisms, the materials affected, factors that influence the rate of damage, appearance or morphology of the damage, prevention and mitigation methods for each mechanism, and recommendations for inspection and monitoring for each damage mechanism.

Corrosion of a metal occurs either by the action of specific substances or by the conjoint action of specific substances and mechanical stresses. Depending upon environmental conditions, corrosion can occur in various forms such as stress corrosion, generalized corrosion, pitting corrosion, embrittlement and cracking. The particular type of corrosion occurring in a specific component can often be difficult to classify. For example, several forms of corrosion (e.g.,

Table 2.3-1. Examples of Corrosion Mechanisms

Mechanism	Description	Types and Conditions
Uniform Attack	Most common type of corrosion and is caused by a chemical or electrochemical reaction that results in the deterioration of the entire exposed surface of a metal.	General attack corrosion accounts for the greatest amount of metal destruction by corrosion, but is considered a safer form of corrosion, due to the fact that it is generally predictable, manageable and often preventable.
Localized	Localized corrosion specifically targets one area of the metal structure.	Localized corrosion is classified as one of three types: 1. Pitting: Pitting results when a small hole, or cavity, forms in the metal, usually as a result of de-passivation of a small area. The deterioration of this small area penetrates the metal and can lead to failure. This form of corrosion is often difficult to detect due to the fact that it is usually relatively small and may be covered and hidden by corrosion-produced compounds 2. Crevice corrosion: Similar to pitting, crevice corrosion occurs at a specific location. This type of corrosion is often associated with a stagnant micro-environment, like those found under gaskets and washers and clamps. Acidic conditions, or a depletion of oxygen in a crevice can lead to crevice corrosion. 3. Filiform corrosion: Occurring under painted or plated surfaces when fluids breaches the coating filiform corrosion begins at small defects in the coating and spreads to cause structural weakness.
Galvanic	Galvanic corrosion, or dissimilar metal corrosion, occurs when two different metals are located together in a corrosive electrolyte.	Three conditions must exist for galvanic corrosion to occur: 1. Electrochemically dissimilar metals must be present 2. The metals must be in electrical contact, and 3. The metals must be exposed to an electrolyte

Table 2.3-1. Examples of Corrosion Mechanisms, continued

Mechanism	Description	Types and Conditions
Environmental Cracking	Environmental cracking is a corrosion process that can result from a combination of environmental conditions affecting the metal.	Chemical, temperature and stress-related conditions can result in the following types of environmental corrosion: 1. Stress Corrosion Cracking (SCC): The combined action of a static tensile stress and corrosion which forms cracks and eventually catastrophic failure of the component. This is specific to a metal material paired with a specific environment. 2. Corrosion fatigue: The combined action of cyclic stresses and a corrosive environment reduce the life of components below that expected by the action of fatigue alone. 3. Hydrogen-induced cracking: Because hydrogen atoms are very small and hydrogen ions even smaller they can penetrate most metals. Hydrogen, by various mechanisms, embrittles a metal especially in areas of high hardness causing blistering or cracking especially in the presence of tensile stresses.
Flow-Assisted	Flow-Assisted Corrosion (FAC), results when a protective layer of oxide on a metal surface is dissolved or removed by the action of fluid flow, sometimes with the added scouring of abrasive particles in the stream, exposing the underlying metal to further corrode and deteriorate.	Some conditions that exacerbate FAC include: 1. Erosion assistance 2. Impingement 3. Cavitation
Fretting	Fretting corrosion occurs as a result of repeated wearing, weight and/or vibration on an uneven, rough surface. Corrosion, resulting in pits and grooves, occurs on the surface.	Fretting corrosion is often found in rotation and impact machinery, bolted assemblies and bearings, as well as to surfaces exposed to vibration during transportation.

Intergranular	Intergranular corrosion is a chemical or electrochemical attack on the grain boundaries of a metal.	This often occurs due to impurities in the metal, which tend to be present in higher contents near grain boundaries. These boundaries can be more vulnerable to corrosion than the bulk of the metal.

Sources: Beginners Guide to Corrosion, Nimmo, William, et. al., NPL, February 2003, Types of Corrosion & Their Effects on Metal, Bell, Terence.

galvanic corrosion, pitting corrosion, hydrogen embrittlement, stress and sulfide corrosion cracking) are characterized by the type of mechanical force to which the metal component is exposed (EC JRC, 2013). Typical elements contributing to elevated corrosion rates in petroleum refineries are presented in Table 2.3-2.

While some of the elements are specific to refinery processes, most are broadly applicable to most industrial activities using equipment and structures made of metal. What is apparent is that to recognize the magnitude of corrosion hazards, one needs to understand the specifics of the corrosive environment and its effects on the materials of construction of equipment and structures. For instance, a knowledge of what conditions cause stress corrosion cracking is needed, before the potential for SCC can be identified in a given process vessel, and a monitoring program initiated.

Another example comes from investigations by the Chemical Safety Board (CSB) of ammonia, refinery, and chemical plants that revealed failures in piping, heat exchangers, and pressure vessels containing hydrogen at elevated temperatures. The mechanism is called high-temperature hydrogen attack (HTHA). This is not the same as hydrogen embrittlement which degrades toughness at low temperatures. HTHA leads to degradation of material properties at elevated operating temperatures and can result in sudden and catastrophic brittle failure.

External corrosion under insulation for vessels and piping systems is of particular concern, because warning signs are not readily visible to the uninitiated from external visual inspection. Once the conditions that contribute to CUI are evaluated based on the materials of construction, process and system operating conditions and environment, inspectors can focus on visual indications

Table 2.3-2. Typical Refinery Elements Contributing to Elevated Corrosion Rates

Refinery Element	Examples
Corrosive substances in feedstock or added or produced in process	Hydrogen chloride, hydrofluoric acid, amines, sulphuric acid, polythionic acids and other Sulphur compounds, oxygen compounds, nitrogen compounds, trace metals, salts carbon dioxide, and naphthenic acids
Refinery processes involving extremes of temperature or velocity	Distillation, desulphurization, catalytic reformers, fluid catalytic cracker, hydrocracker, alkylation
Local conditions	Age of equipment, volume and rate of productions, atmospheric conditions (e.g., climate), planned and unplanned shutdowns

Risk management measures	Frequency of inspection, risk assessment and ranking practices, equipment inventory management, maintenance and repair procedures, auditing and implementation of feedback, use of safety performance indicators

Source: EC EUR 26331 EN – Joint Research Centre Report, Corrosion Related Accidents in Petroleum Refineries

and warning signs of CUI and make recommendations for more in-depth inspection to determine the severity. This includes determining where condition monitoring locations (CML's) should be located to inspect the areas where damage is expected to occur based on the process hazards.

As equipment ages, it becomes more susceptible to corrosion induced failure, in the absence of robust management systems to monitor and counter the corrosive degradation. Baseline metal thickness data need to be collected early in the life of the equipment, so that corrosion trending can be initiated before identifiable metal loss has begun.

2.3.5 Warning Signs

When the physical dimensions of equipment are diminished due to degradation (wear and tear) or corrosion, the ultimate strength will also diminish making the system more vulnerable to physical failure at normal operating conditions. Corroded electrical contacts may impede current flow causing arcing (potential ignition source) and failure to function. Corrosion/erosion of centrifugal pump impellers or casings may result in reduced flow and discharge pressure. As equipment ages, chronic failures may occur with increased frequency. Corroding re-bar within concrete will cause popping of the concrete. These serve as a warning sign of a more significant failure yet to occur.

Some common machinery failures related to metal failure mechanisms indicating problems include:

- Bearing or shaft failure
- Cylinder valve failure or piston ring breakage
- Rotating elements chipping, breaking or distorting
- Scoring of compressor or engine cylinder walls
- Casing distortion
- Gear or coupling breakage
- Seizing of moving parts

Prior to the 1950's, electrical wiring was insulated with hard rubber or non-conducting cloth wrap. Electrical wiring within commercial buildings (including chemical plants and warehouses) utilized a system of knobs and tubes which separated hot leads from combustible materials such as wood. This technology has long since been obsolete. Codes now require conduit or Polyvinyl Chloride (PVC) insulated cable. These are less sensitive to heat and moisture.

Vintage wiring within older buildings is subject to electrical shorts leading to possible power failures or fires. As older insulating materials age, they tend to dry out or split and they may lose their insulating properties. The obvious solution may appear to be a system retrofit or replacement. However, this can introduce additional problems such as exposure to mold or asbestos within walls. A holistic and integrated approach to retrofitting building electrical systems must be considered. Consider all the angles and don't be caught by surprise.

Failure to heed the warning signs in a timely manner can lead to unfortunate outcomes, when failure inevitably occurs. A mid-sized oil refinery in Eastern Canada had undergone a few changes in ownership. When the refinery had been in service for approximately 20 years, it was finally recognized that the fired heater tubes in a naphtha hydrotreater unit were severely corroded and needed replacement. The heater coil operated at a pressure of 600 psig with a coil outlet temperature of approximately 625 °F. In anticipation of a turnaround and an opportunity to re-tube the heater, a contractor crew began mechanical preparations on the furnace structure while the unit was still operating. An operator responding to an upset condition in the unit began adjusting the burners from a catwalk at the side of the heater. Meanwhile, preparations for maintenance continued. A sudden tube failure in the firebox resulted in a large fireball which engulfed the heater killing the operator and a nearby maintenance contractor. The severely corroded tubes had reached the end of their lifecycle and were unable to sustain direct flame impingement which occurred during the upset. Was this simply bad luck, misplaced trust or failure to deal with an identified deficiency in a timely manner? The latter appears to be a contributory cause to the failure. When it comes to the integrity of aged equipment, it does not pay to gamble. Unless there is certainty about the equipment condition, it should be shut down and the aged components replaced.

2.3.6   Aging Equipment Failure Case Studies

An example of aging failure due to metal fatigue is shown in Figure 2.3-2. It shows a rotor from a turbine which has split in two. This example involved a power plant in Germany which experienced a catastrophic mechanical disintegration in a large electrical generator that was in start-up mode.

The incident destroyed the generator and scattered debris over a wide region. One large piece of the rotor shaft was propelled a distance of one kilometer. Operating personnel alarmed by high levels of vibration fortunately sought shelter. Had it not been for early morning on a public holiday, there might have been casualties in the surrounding community. The contained energy "explosion" originated in the central axis of the main rotor and resulted in the total disintegration of the shaft. The generator had only been in service for 16 years

Figure 2.3-2. Catastrophic Failure of Electrical Generator Rotor

and had recorded 58,000 hours of operation. During this period, it had performed 110 cold starts and 728 hot starts. It was also revealed that the rotor was the largest single casting ever produced for this particular service. In other words, this was a one-of-a-kind design. Examination of the fractured shaft revealed signs of crystal inclusion in the grain structure which weakened the material. This situation occurred during manufacture and was not detected through the entire lifecycle. The initial commissioning 16 years earlier had encountered vibration problems. These were not fully addressed and were corrected by counter balancing the shaft. Further evidence revealed a long history of vibration problems during startups. The aging process is impacted by how the equipment is operated as well as to the original process design, including the effects of process upsets.

Figure 2.3-3 depicts another case study reflected by the aging in a railway siding built several decades ago to provide access to nearby chemical plants.

With the recent closure of many facilities, only two small plants remained. As a result, usage of the line has decreased and it is only used occasionally. Many railway ties have split and the rails appear severely rusted. Weather records confirm that the area has sustained serious flooding several times over the past six years. It appears that fresh gravel has recently been dumped onto the line to fill the void between the ties. What you have just observed should drive some serious questions. Is the current line suitable for service? Has the gravel merely masked a deeper problem associated with ground erosion and sinking? Can the line be safely used and, if so, should there be any load and speed restrictions?

Figure 2.3-3. Rotted Railway Ties Providing Weakened Support

And finally, a case study related to aging cable. In the first half of the last century many industrial and commercial buildings were wired with RH-BX cable. This cable, believed to have a life expectancy of 30 to 40 years, was comprised of armored metal conduit over electrical wiring. The internal wires were insulated with a rubberized coating which over time can dry out and crumble off. This can contribute to arcing if a bare conductor comes into contact with a metal service. At the time of initial installation, there was limited experience with electrical wiring in buildings and codes did not specify test measures and/or mandatory replacement.

Some vintage facilities still exist with this obsolete system of wiring. The risk of fire increases significantly when modifications or retrofits are carried out. Attempts to cut or splice RH-BX cable often result in the total destruction and spalling of the insulation within the conduit. The message is quite clear. Replace vintage wiring at the earliest opportunity if a building is to remain in useful service and particularly if the building is occupied by workers on a full time or part time basis. Do not attempt short cuts or interim repairs with older wiring since this could result in early failure.

There is a major lesson to be learned from these case studies. The clock starts ticking on the aging process at the time of initial equipment commissioning. The aging process is impacted by how the equipment is operated as compared to the original process design, including the effect of process upsets. If the equipment has not been properly designed, constructed, installed, operated, maintained and protected, it will deteriorate at a faster rate and it may fail when least expected. Such deficiencies are commonly referred to as "birth defects" that may or may not always be detected through infant mortality (functional) failures. Don't take chances with new equipment. If it does not perform to expectations take the time to investigate why. Understand how equipment operates and correct any deficiencies before commencing operation. There may not be a second chance to get it right.

Metal fatigue is of particular concern in the airline industry. In April 1988, a Boeing 737-200, operated by Aloha Airlines experienced sudden structural failure of the fuselage and a consequent explosive depressurization while on route from Hilo to Honolulu. Approximately 5.5 meters (or 18 feet) of cabin skin and structure was lost from the aircraft and one of the cabin crew was fatally injured. The flight crew carried out an emergency descent and made a landing at Kahului Airport on the Island of Maui, Hawaii.

The following is an excerpt from the official National Transportation Safety Board (NTSB) report (NTSB 1989):

> As the airplane leveled at 24,000 feet, both pilots heard a loud "clap" or "whooshing" sound followed by a wind noise behind them. The first officer stated that debris, including pieces of grey insulation, was floating in the cockpit. The captain observed that the cockpit entry door was missing and that "there was blue sky where the first-class ceiling had been." The captain immediately took over the controls of the airplane. He described the airplane attitude as rolling slightly left and right and that the flight controls felt "loose". This accident was attributed to high stress levels within the aircraft likely the result of many short haul trips.

According to the official accident report, two inspectors working on separate shifts conducted inspection as required by Boeing Service Bulletin (SB) and related Airworthiness Directives (AD) after work that had been done on the aircraft fuselage skin prior the accident. An inadequate maintenance program was found to be the reason for the fuselage section separation during flight. The maintenance program failed to detect the presence of significant disbonding, corrosion, and fatigue damage. The process that was used to bond the overlapping fuselage skins together was poorly performed, and led to early disbonding (NTSB 1989).

The disbonding in turn resulted in what is known as a "knife-edge effect". This effect created a poor fatigue detail in the skin and many adjacent fastener holes started to crack. This form of cracking is known as multiple site damage, which leads to Widespread Fatigue Damage (WFD) in its advanced stage. By definition, WFD is a condition in which the airplane is no longer able to carry the required residual strength loads. Root causes of this incident were failure to understand metal fatigue mechanisms and inadequate inspection by a knowledgeable person after repairing the skin.

2.4   SYSTEM FUNCTIONAL AGING

Functional aging can occur in production and manufacturing support systems, including power supply (e.g., electrical, hydraulic and pneumatic) and electrical, instrumentation and control systems, Safety Instrumented Systems (SIS), and data management systems to name a few. From investigation of major industrial incidents, it is revealed that system functional failures are sometimes initiating events (e.g., power failure, instrument air failure) or more often, contributing events (e.g., safety critical measurement failure, interlock failure).

Since the advent of safe automation standards (ISA 84 and IEC 61508), considerable attention has been focused on the classification (Safety Integrity Level (SIL)), design and testing of Safety Instrumented Systems, and application to Safety Instrumented Functions (SIF). Strictly following the requirements of the standards, particularly the testing, maintenance, and ensuring the reliability of support systems, should lower the failure rate for new installations due to functional aging.

Accelerated functional aging of equipment can be caused by not having designs which support safe maintenance practices, e.g., electrical systems without isolation and separation to maintain partial plant load. Another example is interlocks that can't be functionally tested while in service.

2.4.1   Aging Equipment Failure Mechanisms

There are still legacy systems throughout industry that have not been upgraded to meet the new requirements. Over time, components (e.g., solenoids, valve plugs, relays) in these older installations can deteriorate resulting in functional failure with catastrophic impact.

An explosion occurred during startup of a gas fired aluminum annealing oven causing the entry door, which was large enough to allow a forklift carrying cast aluminum logs to enter, to separate from the oven and land 50 ft. away. Fortunately, no employees were injured at the time. The incident investigation found that the Safety Shutoff Valve (SSV) installed on the burner pilot gas line had allowed gas to leak into the idle oven, and the purge cycle timer had been adjusted (by whom and why was never determined) to shorten the purge time, which allowed an explosive mixture to accumulate prior to light-off. The SSV was removed and disassembled revealing a groove on one side of the plug that caused a functional failure. The groove appeared to have been caused by erosion due to the plug rubbing on a plug guide in the body of the valve. Further investigation of the manufacturer's installation instructions found a warning about not installing the valve with the stem at an angle greater than 15 degrees from vertical. The valve had been installed with the stem at 90 degrees from vertical, which led to erosion of the plug. The oven had been in operation for several years, and had a burner system that complied with National Fire Protection Association (NFPA) standard when installed. Had there been preventive maintenance involving periodic leak testing of gas train SSVs, the functional aging might have been detected before the incident occurred.

Another case involved a large coal fired industrial boiler in an auto plant that sustained an explosion that killed six workers and severely injured 14 others. The vintage boiler, which had been built 60 years earlier was totally destroyed. Workers at the facility had long complained about antiquated equipment and dangerous conditions. At the time of the explosion the boiler had been shut down for annual maintenance. Although workers took measures to stop the flow of gas (gas firing was used to heat the boiler before lighting the coal burners) into the boiler, fuel gas continued to seep in through a faulty isolation valve. More than four hours later the accumulated gas was ignited by hot slag or some other heat source, causing the massive explosion and fire.

Note also, that both of these incidents occurred during non-steady state operation, which is when a majority of incidents occur, and when system functionality is often critical.

2.5   AGING STRUCTURES

Physical structures must be designed to support vertical loads as well as withstand extraordinary forces due to weather and/or other activities. Structural design practices are well established and take into consideration physical dimensions, the mechanics of stability (statics and dynamics) and the strength of various materials. As highlighted earlier there are many types and categories of materials used to construct and assemble process equipment and infrastructure. Some materials define the size and shape of equipment while others serve to secure or enclose. The functions served by various materials are as diverse as the failure mechanisms. We have chosen to start this chapter by examining the many ways in which metallic materials might fail. This assumes that metallic components are present in at least some form in most systems. The order in which this material is developed and presented does not suggest priority. Your own experience or situation may suggest that structural failures involving concrete settling are more significant. The integrity of well-designed structures is always subject to some uncertainty related to ground conditions which may be changing. Another important variable may be the quality of bolts, fasteners and mortar between structural components. Ultimately, a good design must stand the test of time and consider all the things that provide strength and that might fail.

What causes structures to fail? In simple terms, when load stresses exceed those anticipated in design, mechanical failure occurs. With aging structures, however, the capability of the design may have changed or diminished due to mechanisms such as erosion, flooding, corrosion, cracking, drying, chemical decomposition of clays and mortars, ground movement etc. This could be aggravated by adding more weight from new equipment, often as a result of debottlenecking projects. Note that change in weight and structural loads, should be considered in Management of Change (MOC) processes where applicable. Progressive failures cause loads to shift to other parts of structures which may then be required to support a higher share of the load than was anticipated in design. Ultimately, failure will occur.

Failure of a structure can occur from many types of problems. Most of these problems are unique to the type of structure or to the various industries. However, most can be traced to one of four causes.

The first is that the structure is not strong enough to support the load. If the structure or component is not strong enough, catastrophic failure can occur when the overstressed construction reaches a critical stress level. This is essentially a design issue.

An actual example is brittle fracture failure during hydrotesting of a newly installed Fluid Catalytic Cracking (FCC) catalyst regeneration vessel with a normal operating design temperature well above normal ambient conditions. The vessel was constructed of high temperature alloy material with a DTBT temperature about the minimum ambient site conditions. On the day the hydrotest was performed, the ambient temperature had been below the DTBT for some time, and the stresses in the vessel due to the weight of water caused catastrophic failure at the bottom head. Failure of the design to consider the effect of minimum exposure temperature on the DTBT of the construction material resulted in failure.

The second is instability, whether due to geometry, design or material choice, causing the structure to fail from fatigue or corrosion. These types of failure often occur at stress points, such as squared corners or from bolt holes being too close to the material's edge, causing cracks to slowly form and then progress through cyclic loading. Failure generally occurs when the cracks reach a critical length, causing breakage to happen suddenly under normal loading conditions.

The next contributor is from the use of defective materials. The material may have been improperly manufactured, acquired defects during forming, or may have been damaged from prior use. Another cause of failure is from lack of consideration of unexpected problems. Vandalism, sabotage, and natural disasters can all overstress a structure to the point of failure. Improper training of those who use and maintain the construction can also overstress it, leading to potential failures.

The last cause is changing conditions (such as aging) within the structure itself or in the service conditions or physical infrastructure (ground movement). Facilities near salt water exposed to high humidity or those exposed to frigid temperature environments experience accelerated aging from external environmental conditions. Attempting to retrofit an older building to current codes and standards may introduce some unique problems and could lead to collapse during the modification stage. Equipment and infrastructure that is not covered by asset integrity management requirements, is particularly vulnerable due to lack of operating and inspection history

2.5.1   Warning Signs

How long can you maintain an aging structure? There is no simple answer to this question. If defects and signs of aging are addressed as soon as they appear and if the causes are properly analyzed and understood, the life expectancy can be extended. However, if a structure begins to sag noticeably and appears distorted even to an untrained bystander it may be time to abandon it and prepare for demolition. A proper structural analysis should be considered to determine wither it is fit for service.

2.5.2   Aging Structure Case Study

Some of the more dramatic aging equipment failures affect structures, which have a direct effect on the public. In recent years there have been several highway bridge collapses that has resulted in fatalites. One of the most publicized was a bridge crossing the Mississippi River on Interstate I-35 near Minniapolis-St. Paul, Minnesota in 2007, which killed 13 and injured 145.

Figure 2.5-1 shows a grain loading conveyor that collapsed in Louisiana in 2012, closing a key highway along the west bank of the Mississippi River. There was no injury due to the collapse, but the traffic diversion added significantly to commute times due to the limited highway river crossings in this stretch of the river.

According to a company official, a support truss failure caused the collapse, as the load shifted to other members of the structure. One possible aging mechanism was cyclic loading causing metal fatigue in the truss. In between ship departures and arrivals, the load on the structure was much less during idle periods. Vibration during operation may also have contributed to the cycling and generation of micro-cracks. Because photographs of the failed member are not available, the type of failure (ductile or quasi-brittle) is not known. External corrosion due to the high humidity environment may also have accelerated the failure, once other factors were in play.

Unrecognized change of service is a common contributor to structural failures. Seldom do user groups have background knowledge pertaining to how and why the structure was built in the first place. In December 1967, the Silver Bridge over the Ohio River collapsed while it supported rush-hour traffic, resulting in the deaths of 46 people. Two of the victims were never found. The Silver Bridge was an eyebar-chain suspension bridge built in 1928 and named for the color of its aluminum paint. The bridge connected two small communities in

Figure 2.5-1. Grain Loading Conveyor Collapse in Ama, Louisianna

West Virginia and Ohio. Investigation of the wreckage pointed to the cause of thecollapse being the failure of a single eyebar in a suspension chain, due to a small defect 0.1 inch (2.5 mm) deep. Analysis showed that the bridge was carrying much heavier loads than it had originally been designed for and had been poorly maintained. The collapsed bridge was subsequently replaced by a new bridge.

Abandoned and forgotten structures can be accidents waiting to happen. An abandoned steel mill in Sparrows Point, Maryland was being prepared for demolition when the roof suddenly collapsed seriously injuring nine workers. This accident occurred in May 2014, as a work crew was removing asbestos from the inner walls. By the mid-20th century, the Sparrows Point mill was the world's largest steel mill, stretching 4 miles from end to end and employing tens of thousands of workers. It used the traditional open hearth steelmaking method to produce ingots, a labor- and energy-intensive process. Today this technology is obsolete. As a result, maintenance had been deferred on the facility for many years. The facility also had as many as six owners in recent years and lack of commitment to maintaining the facility likely was a major contributing factor.

Before attempting to demolish abandoned structures, a structural evaluation should be performed to determine its asset integrity. If it is deemed too hazardous to perform such an evaluation, then safety measures should be undertaken, such as temporary supports or removal of some portions of the structure to protect working crews from catastrophic failures of overstressed members.

Some examples of aging equipment and structures are shown on the following pages.

Figure 2.5-2. Image of Corroded Oil Recovery Vessels

Figure 2.5-3. Image of Aging Iron Making Facility

Figure 2.5-4. Image of Aging Gas Plant

Figure 2.5-5. Image of Aging Process at Marine Facility

Figure 2.5-6. Image of Aging Process Facility

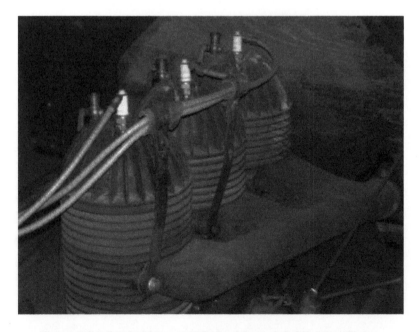

Figure 2.5-7. 1911 Vintage 3-Cylinder Internal Combustion Engine

3

PLANT MANAGEMENT COMMITMENT AND RESPONSIBILITY

3.1 PROMOTING SITE SAFETY CULTURE

It can't be said too many times, that a plant's safety culture is driven by the leadership from the top. This responsibility falls most directly on the plant or site manager and his or her direct reports. The front-line workers need to believe that the plant management takes the issue of aging equipment hazards seriously by showing demonstrated commitment to a plan to address problems on an ongoing basis. As in most endeavors, deeds speak louder than words. Many volumes have been written on how to attain an exemplary safety culture (e.g., CCPS 2005 *Building Process Safety Culture: Tools to Enhance Process Safety Performance*, CCPS 2007 *Guidelines for Risk Based Process Safety*), which can be a resource for continuous improvement. It is also important that safety culture is extended to the area of managing aging equipment and infrastructure, which is easily ignored. However, when there is a serious incident involving aged equipment, ignorance on the part of the plant manager will not be an adequate justification.

Division or corporate management also has a share of the responsibility for promoting process safety culture. This responsibility often reveals itself in the approval of budgets for facility expenditures. In the case of aging equipment, this would typically include maintenance, inspection, and decommissioning budgets. Understanding the risks of aging equipment and infrastructure is necessary for good management decision making. Committed plant managers, when submitting cost budgets, may assist in this understanding by explaining the risks of allowing aging or unproductive assets to continue in their present state. This book provides many case studies that may be useful in making a persuasive case for committing resources for repair, replacement or removal of old assets.

3.2 MANAGEMENT CHALLENGES

Plant and facility management has some unique challenges related to the continued operation of aging facilities. As equipment ages, it may be more prone to failure and unexpected downtime. Production targets will need to be adjusted and communicated clearly to workers on the front line. There is often a tendency to make up lost production by increasing production throughput. This may be unwise and unsafe given some of the uncertainties in dealing with aging equipment. Strict adherence to operating guidelines is always important but especially so when the facility approaches the end of its lifecycle. This is the essence of the Risk Based Process Safety (RBPS) element "Conduct of Operations".

Aging facilities and equipment in a state of disrepair are often modified or re-rated. This can apply to infrastructure systems such as steam generation and fuel distribution, as well as process equipment. When this happens, the design (nameplate) operating conditions will need to change. Old habits die hard and tenured operating staff might view the operation as they have in the past. To deal with this, the previous operating history (e.g., capacity, load and specified conditions) will need to be erased from the memory of seasoned workers. New procedures and safe operating limits will need to be established with a technically defendable rationale. This in itself is a communication challenge and a classic example of "Operations MOC".

Used equipment is frequently employed in capital project work especially when a plant reaches full maturity. It is important to recognize that although used equipment can save money on project costs, such equipment still needs to be inspected and fitness for use determinations made prior to use. The maintenance program will need to accommodate this used equipment and funding must be provided.

At a higher level, corporate management may be placing pressure on a facility to produce more while reducing cost. It is unlikely that corporate management is physically located near a production facility or has an intimate knowledge of plant vulnerability in an aging situation. Local plant management is faced with resisting such pressure while attempting to educate those at a higher level of the risks. This is perhaps the biggest management challenge.

3.3   MONITORING AGING PROCESS AND MEASURING PERFORMANCE

Managers of older plants with assets approaching or exceeding their intended lifecycle, should be aware of potential warning signs. Is there considerable Out of Service (OOS) and decommissioned equipment at the site? Is equipment reliability, based on work order and inspect/repair history, at the tail end of the bathtub curve? Is the maintenance department increasingly dependent on used or scavenged parts for making repairs, due to large or growing backlog of reactive maintenance work, or unavailability of parts due to obsolescence? Have trending results for the asset integrity program revealed an increase in damage impacted remaining life?

Plant managers who have been assigned legacy facilities that were recently acquired from other companies need to be keenly sensitive and wary of the conditions of the aging assets they inherited. Maintenance and inspection records should be obtained so that data is available to support ongoing maintenance and inspection decisions. There is a need for creating awareness in and among upper management of the issues they need to consider with respect to process safety during the merger and acquisition process. Whether merger and acquisition are under consideration or not, senior management should investigate to determine if serious problems exist with older equipment and infrastructure, before it is too late.

When signals or situations such as these are evident, the manager should initiate inquiries about the condition of the asset. The answers to the following list of relevant questions can assist in deciding future actions.

- When was the asset installed?
- What was the manufacturer's or supplier's lifecycle estimate?
- What do inspection reports tell about its current condition?
- For metallic components, are corrosion rates/mechanisms known and understood?

- Is the asset fit for service? Has Fitness for Service (FFS) been evaluated?
- What is the estimate remaining service life today?
- Has the maintenance archive indicated what equipment and parts have been replaced and what repairs have taken place over the lifecycle of the facility?
- Are process incident archives available for this asset?

Other metrics should be devised to monitor the health of the asset integrity program implementation. For example, overdue scheduled inspections as a percentage of total inspections required may indicate inspectors are overloaded, or the program is not appropriately focused on the higher risk situations. Another example is comparing the submission dates for inspection reports requiring further actions, with work order completion dates, to determine whether actions are performed in a timely manner, allowing for equipment access. Some examples of leading and lagging metrics are provided in Table 3.2-1.

Table 3.2-1. Component Condition Health Metrics

Metric No.	Indicator	Scoring Element	Definition
1	Leading	Service Age / History	Percent of component age compared to expected life of component adjusted for the remaining life based on current operating conditions including the impacts of normal and upset operations
2	Leading	Obsolete Equipment	Component make and model matches equipment on obsolescence list. Equipment obsolescence is defined as the state where equipment may be difficult to maintain, the vendor no longer supports the product, spare parts are no longer available, or equipment parts become incompatible.
3	Leading	Problem Equipment	Component make and model matches equipment on problem equipment list. This metric represents the identification of equipment where undesirable functional or operational issues have been detected which is suspected to be or is a direct result of a manufacturing defect or in-service configuration with system-wide implications.
4	Leading	Physical Condition	Assessment of component from inspection, testing and maintenance program results based on the asset integrity program

5	Lagging	Functional Performance	Assessment of component performance based on review of maintenance and operations history against performance criteria.

Table 3.2-1. Component Condition Health Metrics, continued

Metric No.	Indicator	Scoring Element	Definition
6	Leading	Operational Efficiency	Measure of operational efficiency based on review of maintenance hours spent on component over past three years compared to target efficiency criteria.
7	Leading	Engineered Maintenance Strategy	Component included in maintenance database (Computerized Maintenance Management System (CMMS))) with defined maintenance strategy (preventive, predictive, condition based, risk based maintenance or maintenance for cause).
8	Lagging	Corrective Maintenance Tasks	Number of reactive maintenance tasks compared to equipment with defined maintenance strategy, excluding maintenance for cause strategy.
9	Lagging	Planned Maintenance Tasks Overdue	Occurrences (count) of preventive maintenance tasks overdue greater than targeted number of days.
10	Lagging	Percent Corrective Maintenance vs. Total Maintenance	Percent of work hours associated with reactive maintenance against the total work hours on the component.
11	Leading	Number of inspectors/maintenance employees holding each type of required certification	A decline in this metric may be a leading indicator of skill gaps or a higher than acceptable backlog for inspection, testing and preventive maintenance tasks.

3.4   HUMAN RESOURCES REQUIREMENTS

Equally important with the monitoring program is the quality and skills of the team members tasked with the asset integrity program. There are some striking examples of companies with best in class engineering standards, inspection practices and evaluation procedures, which only achieve substandard asset integrity results due to failure to address human resource issues.

For example, one company assigned recently graduated mechanical engineers to the inspection department, with the responsibility to review inspection reports submitted by field examiners to determine further action. While examiners do record most observed conditions, they do not always indicate a need for follow-up, especially if the conditions are not deemed deficient. Corrosion under insulation is an example, where knowledge of some specific warning signs is needed to appreciate what might be happening out of sight. Because of the practice of using inexperienced, non-certified (against API standard and company policy) personnel with a high turnover rate (they moved on to other assignments), many large columns experienced severe CUI after many years of operation, with some not passing a FFS evaluation and having to be replaced. In this case, the CUI issues were not identified due to inadequate training and experience of personnel that were intended to stay in that position long enough to gain necessary skills.

Certified inspectors, field examiners and data analysts that apply American Petroleum Institute (API) inspection practices have the experience and authority to extend the scope of an inspection to determine whether there is a need to do additional investigation (e.g., remove sections of insulation) based on known warning signs.

Understanding the asset integrity program needs, and matching it with the appropriate skills, is essential for obtaining quality of implementation. Performing a skills assessment can reveal whether the team has the proper skills. Do personnel who are monitoring Non-Destructive Examination (NDE) contractors have the necessary certification/qualifications on all techniques to interpret results? What is the training and experience level of the mechanical engineers responsible for pressure vessel and piping integrity? Are there Subject Matter Experts (SMEs) at the facility or in the company? Are they consulted and is their advice followed? Do maintenance technicians have adequate skills to work on large complex equipment (e.g., centrifugal or reciprocating compressors)? What skills are maintained in-house and what should be outsourced? If skills are outsourced, does the industry have sufficient asset specific expertise to effectively monitor their performance?

The gaps found from the skills assessment should be used to develop or update a skills training plan for maintenance and inspection personnel. The plan should identify what skills require ongoing refresher training such as certifications (e.g., welding, inspection). For maintenance technicians, periodic scheduling of rotating equipment vendor workshops might be appropriate.

3.5 PLANNING FOR EQUIPMENT RETIREMENT AND REPLACEMENT

Process plants and infrastructure are typically designed with a specified life expectancy. Traditionally, upstream petroleum production facilities were often designed for shorter life expectancies, to align with the production decline rate of the field. Due to technology advancements and oil price increases, some of these facilities are now operating well beyond the original life expectancy of the field. Most downstream process industry facilities were designed for longer life expectancies, generally 30 years plus. While the life-cycle may be longer in some cases, eventually aging assets will approach a time when it may become necessary for safety and business reasons to overhaul, replace or retire them. Managing the asset at this point in the lifecycle may require more commitment of resources to monitor conditions, and further capital expenditures may be required to upgrade the asset.

Chapters 1 and 2 describe some of the mechanisms for aging of assets. As the asset nears the end of its initial lifecycle target, more attention to monitoring conditions like corrosion or cycling fatigue should be expected. Appropriate budgeting for inspection, testing and addressing asset integrity concerns of aging assets needs to be provided.

Managers, Engineers and Business Development teams need to appreciate that the day will come when equipment has reached the end of its usable life and a capital plan needs to be developed to manage this occurrence, whether it be replacement or retirement. Running equipment with known deficiencies at the end of remaining life is not acceptable when dealing with hazardous processes given the potential for failure resulting in extensive property damage and business interruption. Proper communication of the capital plan to executives with financial responsibility is needed, to properly develop and implement a repair / replacement strategy so that the equipment at end of life does not fail before the replacement equipment can be put into service. This includes business continuity planning and equipment contingency planning to reduce process downtime. This is to avoid surprising executives that a major capital project may be necessary. Like everyone else, executives do not like to be blindsided when it comes to major capital expenditures that they were not adequately prepared. The capital plan for dealing with the aged asset should be communicated well in advance of the scheduled date for implementation.

Strategic Planning is a tool that can be applied to this situation. Strategic planning is an organization's process for defining its overall business direction, and making decisions on allocating its resources to pursue its long-term business objectives. It may also extend to control mechanisms for guiding the implementation of the strategy. Strategy has many definitions, but generally involves setting goals, determining actions to achieve the goals, and mobilizing resources to execute the actions. A strategy describes how the ends (goals) will be achieved by the means (resources). The senior leadership of an organization is generally tasked with determining strategy.

Because aging assets are often physically large (major rotating and/or fixed equipment) or extensive (steam distribution infrastructure) and outside the normal run/maintain budget cycle, having a strategy for managing these non-routine expenditures makes good business sense (see CCPS booklet *Making the Business Case for Process Safety*). It needs to begin at the top. Strategic planning will never succeed if leaders delegate it, because it requires their understanding and buy-in. Getting buy-in and commitment at all levels helps mobilize and gives everyone a stake in the endeavor to make the plan a reality.

A thorough understanding of the expertise needed to address dealing with assets that have reached end-of-life is paramount. Without this, management may not be asking the right questions or may not get the right answers. Engage a diverse team with diverse skills sets including operators, asset integrity personnel, process engineers / designers and plant management to fully understand the hazards the process creates for the equipment, operating conditions/environment, equipment conditions/performance/environment, history and asset integrity program trending results (including deficiency management/ repairs/replacement of component) to adequately evaluate the equipment risk. The output should include guidelines for making decisions at all levels. Think of the strategic plan for aging assets as a guidebook for any major capital decision the organization may encounter. It should be one of the most important tools in a manager's toolbox. It should also anticipate and manage change. With aging assets, change is inevitable. While the strategic plan is a guideline, it needs to be flexible enough to respond to an ever-evolving conditions and situations.

3.6 APPRECIATING THE IMPORTANCE OF AGING INFRASTRUCTURE TO THE BUSINESS ENTERPRISE

Senior management need to appreciate the significance of Aging Infrastructure. They need to understand that infrastructure does not last forever and is as important as process equipment, since a failure in a utility and/or support system can result in a major process upset and emergency shutdown of the process. This can lead to physical damage for the equipment and the product with extended business interruption. Infrastructure failures can lead to unsafe and unreliable operations and may severely impact in the bottom line. Some examples of possible issues caused by aging infrastructure are listed below.

3.6.1 Structural Assets

- Aging structural assets such as buildings can create issues with water leaks and mold due to moisture ingress, deteriorated water lines and HVAC systems. Buildings may no longer be suitable for "shelter-in-place" or "safe heaven". One company had to abandon and demolish its office buildings because maintenance and clean-up costs became too high. Personnel were relocated to rented office space.
- Deteriorating pipe rack supports or fire proofing on support steel can cause or contribute to pipe rack failure and major releases of flammable or hazardous materials. According to a CSB investigation report the severity of the fire at a Valero Refinery in February 2007 was partially caused by the sudden collapse of a pipe rack due to lack of fireproofing.

3.6.2 Roads

- Poorly maintained roads can cause mobile equipment damage and increase the risk of serious incidents including occupational injuries.
- Uneven surface in trailer storage areas can cause trailers to tip over with possibility of chemical release and injury.
- Improper site grading can lead to severe ground erosion as a result of flooding.

3.6.3 Impoundments and Dikes

- Deteriorating impoundments may cause ground or water contamination with high clean-up costs.
- Leaking dikes will not only cause possible ground contamination but will allow hazardous material to spread increasing the area of exposure to flammable and hazardous materials releases.

3.6.4   Fire Water, Cooling Water and Sewers

- Deposits collecting over time can reduce flow capacity.
- Firewater system failure can cause operational issues resulting in a lack of sprinkler protection (may violate local authority or company right to operate) and prevent maintenance requiring hot work.
- Loss of the fire protection system exposes the protected assets to fire damage in the event of a fire. A process shutdown (until the system is restored), may be required to reduce the fire exposure.
- Process Sewer failures can cause environmental contamination and require large clean-up cost if not detected quickly.
- Loss of cooling water may force the shutdown of the process or cause a possible release from the overpressure scenario.

3.6.5   Electrical Distribution Systems

- Faults in inadequately designed / calibrated electrical distribution systems can result in cascading electrical failures resulting in upset conditions, shutdown of affected processes, physical damage to equipment and business interruption.
- Poorly maintained electrical distribution systems can cause equipment failures and un-planned shutdowns causing physical damage and loss of production.
- Faulty electrical distribution equipment can also be the ignition source in an area handling flammables or cause electric shock hazards for personnel.
- Inadequate emergency power / uninterruptible power supplies for critical processes can adversely impact continuity of operations in the event of a loss or normal electrical service.

3.6.6   Marine Facilities

- Possible over water and subsea leaks causing spills and water contamination with additional clean-up cost and regulatory exposure.
- Deterioration of docks and piers can impact the ability to ship and receive raw materials and finished products.

3.6.7   Other Process Facility Infrastructure

Some additional examples of infrastructure where service impairment or failure due to aging can have significant impact on operation and safety:

- Plant air and nitrogen supply
- Heat transfer fluid systems
- Waste stream treatment including waste water treatment and process gas treatment (thermal oxidizers)

In general, these infrastructure failures can increase operating costs, cause loss of production affecting customers and sales contracts, contribute to incidents increasing the workman's compensation costs, increase environmental liability and clean-up cost and expose the company to lawsuits and regulatory fines.

3.7 ADDRESSING AGING INFRASTRUCTURE IN DECISION PROCESS

Plant Management needs corporate support to properly maintain and upgrade aging infrastructure. The capital required for upgrading aging infrastructure is hard to justify since it typically does not meet the corporate "Return on Investment" guidelines for capital spending. It is up to senior management to assure that infrastructures issues are brought to their attention and appropriate funding is provided so that equipment is repaired or replaced in a timely fashion before reaching end of life and failure in service.

Resource requirements also need to be addressed by senior management. Inspectors and maintenance personnel need to be trained to be available to identify infrastructure issues, recommend corrective actions and oversee implementation.

Senior management needs to create a process safety culture that places value in infrastructure. It is up to the corporate leadership to establish site management goals and key performance indicators that include managing aging infrastructure.

Executives and Plant Management should require each site to identify and prioritize their process safety risks and include infrastructure issues as part of this evaluation. Insurance audit findings can be a good source of information during these reviews. Risk-Based Methodologies should be used for prioritization of improvements as seen in Chapter 4. Corporate Risk Tolerance Guidelines should include reference to infrastructure issues and how they should be rated relative to other risks.

3.7.1 Questions Executives Need to Ask

The following questions should be reviewed at the highest level of the organization to assure that they are included in the business strategy and annual operating plan discussions (refer to ACC Responsible Care® Process Safety Code of Management Practices)"

- What is the remaining life of critical infrastructure assets?
- How was the risk for aging infrastructures issues determined?
- What are the consequences of infrastructure failure?
- Which deficiencies carry the highest risk?
- Do the human resources have the proper competency?
- What inspections are best performed by outside resources?
- What is the cost of maintaining or upgrading the infrastructure and correcting deficiencies?
- What time frame is required for completing the repairs or replacement? 5 to 10-year plan?
- What is the business continuity plan in the event of an infrastructure and/or equipment failure?
- What is the equipment contingency plan, including sparing options for aging equipment to reduce the downtime in the event of a breakdown?

The corporate management commitment to infrastructure improvements has to be evident by proper financial and manpower support to the sites. It is up to the corporate leadership to establish goals and key performance indicators that include managing aging infrastructure issues. Maintenance of infrastructure should be part of site mangers goals.

Senior management should establish and review on an established frequency key performance indicators such as:

- Percent of expense budget spent on infrastructure repairs/upgrades
- Number of infrastructure inspections completed
- Number of infrastructure deficiencies open
- Amount of capital spent on infrastructure upgrades

3.7.2 Mergers and Acquisitions

During mergers and acquisitions senior managers need to instruct the due diligence team to audit the condition of infrastructure and include cost estimates for repair in integration costs. The team should review insurance and Environmental Health and Safety (EHS) audit findings to identify aging infrastructure. Incident reports can be helpful in identifying issues with underground lines and utilities. Unidentified Infrastructure problems can be very costly and could affect the bottom line of the acquisition or merger.

4

RISK BASED DECISIONS

The use of risk analysis methodologies for managing process safety and risk has been in existence for some time. Methodologies such as Fault Tree Analysis (FTA) and Failure Modes Effects and Consequence Analysis (FMECA) were commonly employed by the nuclear and aerospace industries for many years prior to the promulgation of common process safety regulations. Over the intervening years, many books and articles have been published and delivered on the use of risk based methods for assisting managers in making decisions that related to process safety management. The reader is guided to a partial list of references at the end of this chapter for additional reading. The purpose of this chapter is to briefly highlight some of the risk-based methodologies that are particularly applicable for decision making regarding aging assets.

4.1   RISK MANAGEMENT BASICS

Risk is sometimes referred to as the science of uncertainty. Simple risk has two components, impact and probability. An exposure term is sometimes used to align the loss with the initiating failure. Some losses do not occur unless the failure occurs at a specific time or during a critical activity. To understand risk is to understand all the possible consequences of failure and their associated probabilities. The total of all such combinations defines system risk. Risk quantifies the level of concern that should be directed towards a problem or issue and helps to strategize and motivate follow-up action. Recognizing that any situation or event may have several possible outcomes, risk typically examines the undesirable or negative consequences of failure. There may be positive effects or gains and these need to be considered especially in the context of business as shown in Figure 4.1-1. Risk decisions are made in the face of uncertainty and, as such, require a protocol or platform to assure consistency. Risk management provides that platform. The diagram below is a simple strategic roadmap for risk management (Kelly, 1998).

The risk management process examines all risks within a defined system and strategizes controls for dealing with them on the basis of the threat they pose. From a broader business perspective, risk management is defined as a process which combines the results of risk assessment with economic, political, legal and ethical considerations to make decisions. Risk decisions need to focus on strategies which maximize gain and minimize loss. Risk management should utilize a lifecycle approach for all systems whether they are physical or non-physical. It begins with a thorough knowledge of and understanding of business goals and objectives. This is the foundation upon which success or failure are measured. The process is continuous and self-sustaining. System monitoring tracks performance and behavior with time against some standard. Anything unusual or out of the ordinary identifies the possibility that the system may be at risk. An activity may be producing results that are unintended and unacceptable. Left unchecked, things may only get worse. System monitoring establishes a normal database against which future deviations may be tracked.

Risk assessment is a strategy used for quantifying and ranking risks. Completed risk assessments provide a great deal of information which can be used to search out future failures and opportunities. It is really the heart of the risk management

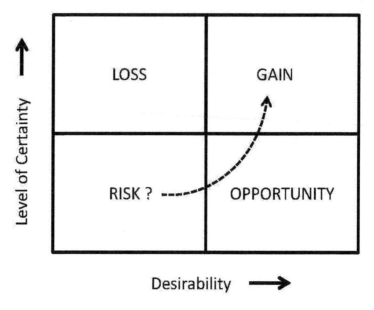

Figure 4.1-1. Dimensions of Choice

process. While many risks are obvious and will be dealt with as encountered through hazard or risk identification, others need to be strategized and dealt with on a prioritized basis. Risk assessment explores the many consequence/probability combinations which exist for every possible failure scenario. It is important to recognize that while most of these represent low risks, when considered collectively, they might indicate a high system risk.

There are many established methods for conducting risk assessment. The methods generally fall into one of four categories (1) Experience based with use of subject matter experts, (2) Relative assessment models, (3) Scenario-based models, or (4) Probabilistic models. Some of these rely heavily on scientific data that may not always be available. Others require the support of trained professionals or external consultants. The most effective form of risk assessment is one that engages a company's own employees and taps into their own experience with operating systems. While such an approach may compromise scientific accuracy, it allows for a broader application of risk management principles across an operating organization.

At a somewhat higher level, a collaborative process that engages experts and people with a broad range of experience is required. Such a process can generate consistency across an organization and eventually establish a business risk culture. This is a semi-quantitative method that relates all risk exposures back to established corporate risk criteria. This important reference check provides the impetus to develop risk controls or alternatively to accept the risks. The risk matrix tries to strike a balance between different categories of loss. For example, safety related losses such as serious injuries may be ranked alongside very high financial losses. In this way, safety will become a realistic business priority. It is important that management fully understand the risk acceptance criteria and endorse it to make the risk management system effective. More information on this topic will be provided later.

On occasion, it may be necessary to conduct a more detailed analysis of some aspect of a risk in order to validate its magnitude. Business and scientific models may be used to predict the outcome of potential failures. External consultants may be retained to provide additional data or provide an external perspective. These additional activities must be coordinated within a company's risk management framework otherwise, competing approaches will be seen to generate confusion. The final product of a risk assessment should be a concise set of conclusions and a ranking of significant risks.

While risk assessment determines consequence and associated probability for several exposures, the terms are not always of equal significance in determining the risk. Within any organization, experienced people tend to be more conversant with the consequence side of the risk equation. Probability should be considered a secondary variable and should be used primarily to rank significant risks. Risk assessment tries to gain an understanding of hazards and situations that can result in loss. No "high consequence" risk should ever be discounted without at least a contingency plan to deal with its occurrence. This includes low probability risks which some persons might consider remote.

When significant warning signs appear, that should trigger a call for additional assessment of the condition and safety of the asset. Like process safety management, this can take the form of an Asset Hazard Analysis (AHA) using one of the simplified methodologies like What-if or Checklist to conduct a brainstorming session with personnel with knowledge of the asset under review. The analysis should be risk based considering the consequences of failure and the likelihood using a risk ranking tool (e.g., a matrix).

4.1.1 Risk Ranking

Risk ranking is a common practice for prioritizing and making risk based decisions without conducting quantitative risk analysis. One should consider risk prioritization for aging equipment separate from other risks that are identified, due to the potential end of life issues. The basis for risk ranking is the risk matrix that has both a consequence and frequency axis. An illustrative example of a risk matrix is presented in Figure 4.1-2.

This risk ranking matrix is best applied for qualitative risk judgments, and it should be used in a similar approach as when conducting hazard identifications. The product of consequence and frequency provides a measure of risk. Each consequence/frequency pair on the risk matrix is assigned a risk ranking that includes risk levels that are tolerable and others that are intolerable. The intolerable risk levels may be further divided into higher and lower risks to prioritize mitigation actions.

The process for developing a risk matrix is to start with the ranges of consequences of concern and then to determine the tolerability level for each. Generally, the most severe consequence range includes one or more fatalities. However, some companies prefer to define multiple fatality events as the most severe range and a less severe range that typically is limited to a single fatality. Some companies also treat offsite or public impacts as more severe than onsite impacts. One argument for the latter approach is that onsite employees have voluntarily, implicitly accepted some level of risk by working in that environment, and are more prepared to handle the consequences of an incident than the general public who are completely unaware of the hazard. Another argument is

			Hazard Severity				
			Negligible (1)	Slight (2)	Moderate (3)	High (4)	Very high (5)
LIKELIHOOD OF OCCURRENCE	Very Unlikely	(A)	LOW	LOW	LOW	LOW	MEDIUM
	Unlikely	(B)	LOW	LOW	LOW	MEDIUM	MEDIUM
	Possible	(C)	LOW	LOW	MEDIUM	MEDIUM	HIGH
	Likely	(D)	LOW	MEDIUM	MEDIUM	HIGH	HIGH
	Very Likely	(E)	LOW	MEDIUM	HIGH	HIGH	HIGH

Figure 4.1-2. Example of a Risk Matrix

that offsite impacts have more far reaching implications in terms of the Business Case for Process Safety and the License to Operate.

There is considerable data on fatality risk tolerability for individuals (CCPS, 2001). Benchmarking data from chemical/petrochemical companies indicates the results depend on whether they are expressed as impact criteria or event criteria. An impact scenario considers all events that need to occur in order to realize an undesired impact such as an injury. An impact scenario will consist of an initiating event and any number of enabling events, conditional events (probability of ignition, probability of personnel in affected area, probability of realizing undesired consequences) and safeguards. An event scenario considers only those events necessary to have a release or condition with the potential for an injury. An event scenario will consist of the initiating event and any safeguards. Event scenarios are typically used for hazard identification whereas impact scenarios are typically used for Layer of Protection Analysis (LOPA) and facility siting.

Tolerable impact criteria for events with the potential for one or more fatalities range from 10^{-5} to 10^{-6} per year, whereas comparable tolerable event criteria range from 10^{-4} to 10^{-5} per year. This implies that typically tolerable event criteria are set an order of magnitude higher than the equivalent impact criteria. This seems reasonable and conservative given all of the additional conditional probabilities that need to be included in determining the frequency of impacts from scenarios.

Therefore, if a company wants to be in the norm regarding their risk tolerability they would choose a tolerable fatality event frequency of $<10^{-4}$ per year or a tolerable fatality impact frequency of $<10^{-5}$ per year. Once the most severe consequence category has been determined and its risk tolerability defined, the same process is used for each of the other consequence categories allowing orders of magnitude between category frequencies.

The risk matrix should also indicate the region (green) where the risk (denoted by consequence and frequency) is considered tolerable or reduced to As Low as Reasonably Practicable (ALARP). Once the consequence severity is established, (which considers the property damage and business interruption exposure based on the scenario) this will define the tolerability target frequency for the

4.1.2 Risk Mitigation Controls

Risk mitigation controls are strategies that reduce risk to a more tolerable level. These may include safety hardware, protective barriers, procedures, regulations, personnel training, business continuity and equipment, as well as contingency planning. Monitoring itself is a risk control in that it gauges the magnitude of an established risk and triggers the need to respond. Compliance audits and inspections also play a similar role. Risk controls should be specifically tailored to the risks which they address. While it may be tempting to apply generic risk controls to all situations, risk cannot be effectively managed without a thorough understanding of both impact and probability. A risk assessment is therefore a pre-requisite to developing effective risk controls.

There are many types of risk mitigation controls and it is often desirable to apply a combination of these. A coordinated program will ensure that risk controls are compatible and mutually supportive. Generally, four approaches are used to manage risk.

1. *Treat the risk.* Given the results of a risk assessment, something may be required to alter either the potential consequence or the frequency (or both) of a system failure.
2. *Tolerate the risk.* Given the fact that a risk is low and there are few alternatives for dealing with it, it may be acceptable to live with the risk. Monitoring can help to determine if and when further action is needed.
3. *Terminate the risk.* If the risk exceeds acceptable criteria and no controls are deemed viable, the situation is rejected outright. This may involve shutting down a facility or closing a business.
4. *Transfer the risk.* Risk substitution allows for a known quantity to be sacrificed in order to protect a critical or highly sensitive system. Risk transfer includes the purchase of insurance, lease versus buy options etc.

To bring closure to the risk management process, it is necessary to test the effectiveness of risk controls. The risk assessment should be repeated using the assumptions made in implementing risk controls. If these risk controls perform as intended, is there any net reduction in significant risks? Better still, is the total system risk reduced? When all significant risks in a system are addressed with controls, it is necessary to repeat the first step and monitor the controls themselves. This will ensure that they meet the intent and are not themselves subject to failure.

4.2   RISK BASED DECISIONS

Risk decisions are strategies devised with due consideration of apparent risk. Risk decisions seldom involve a simple choice between two options. They require detailed knowledge and an understanding of the things that can go wrong even if the best strategy is adopted. Often the best risk decision is the one with the fewest potential adverse consequences.

How are risk decisions made and what are the ramifications? Risk decisions must consider all potential scenarios with an undesirable outcome. This should include not only physical failures but economic and social anomalies as well as abnormal behavior of people. For example, a unit of aged equipment left abandoned in an obscure location may attract curious onlookers or scavengers who could incur injury from contact with an unknown entity. Informed risk decisions need to exhibit the following:

- Reference to and utilization of hard historical data if such exists
- Adherence to a prescribed risk methodology
- Conformance to corporate risk tolerance criteria

Ad hoc risk decisions void of the above considerations, present a high level of uncertainty and may suggest negligence if a serious incident were to occur. Risk decisions affecting aging equipment and infrastructure should be defendable.

A couple of household examples might help to illustrate risk-based decisions as they pertain to aging. Most shingled house roofs have a life expectancy approaching 25 years. After two decades of harsh weather, shingles tend to warp and split. It should be obvious to anyone at ground level that there might be a problem. Left unattended, serious damage could result to the home. This problem presents four options:

- Repair the damaged shingles
- Ignore the problem until all the shingles are damaged and need replacement
- Ignore the problem until physical damage is apparent within the home
- Replace all the shingles even though damage to the roof is not yet apparent

The first option sets up a never-ending pursuit of catching the next defect before it escalates to a major problem. This is essentially the Band-Aid approach. How does one keep track of old versus new shingles? How do you manage quality and color differences between old and new? How do you replace some shingles without damaging those adjacent to them? Finally, the cost of piecemeal repairs may greatly exceed the cost of doing the job right in the first place.

The second option simply ignores the problem and accepts the risk blindly and head on. When the roof finally reaches a state of "high disrepair" there may be significant damage (mold and rot, cracked joists) to the roof membranes or trusses beneath. A major cost may be incurred.

The third decision option is a more aggressive approach to option 2. It not only accepts the risk but the certainty of damage. The last option is the preferred choice by many. It might seem somewhat conservative but it recognizes that shingle replacement is ultimately a necessity. This decision option merely advances an expenditure that will be required in a few years anyway. It buys peace of mind and an assurance that the roof should not leak for many years.

A similar situation may exist with a household hot water heater. Tank failures tend to occur frequently as they approach twenty years in service. Sediment in the water is a major contributor to overheating of the tank shell. Often such failures occur with little warning and may cause significant water damage to a home. Unlike the example cited with the shingled roof there is no practical way to inspect a tank without dismantling it and cutting it open. The risk decision in this case is somewhat simpler. Replace it at a nominal cost in the fifteen to twenty years' life or risk a failure with consequential damage. In both cases, a tank replacement is required. The preferred option should be obvious.

Applying risk based decision techniques can guide the decision maker to making an informed and sound determination of the appropriate strategy. The first step is to determine the extent and understand the consequences of failure of the equipment. In this regard, there can be several failure modes (e.g., from leakage or gradual loss of functionality, to sudden catastrophic failure) and multiple outcomes (e.g., odor, frequent outage, to major fire/explosion, loss of life, public evacuation). A proper evaluation should consider several scenarios because the worst consequence case scenario is typically also the least likely, hence events with lesser severity outcomes can pose higher risk if the frequency is much higher.

As previously indicated, risk has two dimensions, frequency being the other. In general, consequences are easier to picture and comprehend. Frequencies are more abstract, especially when the event repeat interval is greater that the life of the plant. Since cost is an objective in most endeavors, rational risk based decisions require assigning probability to the hazard scenarios. This shall be further discussed later.

4.2.1   When to Apply Risk Based Decisions

Not all decisions involving aging equipment require in depth risk based decision analysis. In general, the more severe the potential consequences of no action, the more benefit afforded by risk based decisions. Criteria for deciding when RBD is appropriate are a matter of individual company risk aversion and financial loss acceptance. However, some generally accepted guidelines for when further quantification of risk is desirable are listed in Table 4.2-1. The criterion for financial loss is highly dependent on the company's risk appetite. This can be influenced by the size/financial position of the company and its ability to recover from an event, which gets to business continuity planning. Even for large companies, the limit at a given plant or site is likely to be $1-5 million.

4.3   HOW TO APPLY RISKED BASED DECISIONS

The steps for how to apply RBD making are outlined in Figure 4.3-1.

Table 4.2-1. Guidelines for Risk Based Decisions

Type	Criteria
Risk	Hazard scenarios determined to be in the Red (high risk) cells of the risk ranking matrix.
Consequence	1. An event with onsite or offsite life threatening potential. 2. Combined business interruption and property damage >$ XXX. 3. Irreversible or > 5-year recovery environmental damage.

	4. Public impact requiring hospitalization or evacuation. 5. National news media coverage, potential damaged company reputation.

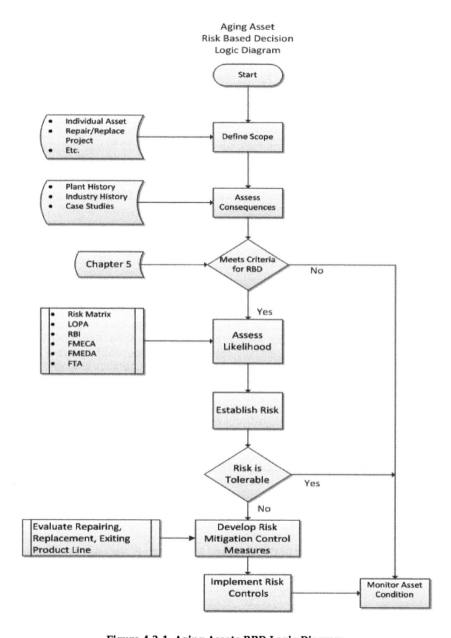

Figure 4.3-1. Aging Assets RBD Logic Diagram

The process begins with defining the scope of the decision action involved. This can range from a single large item such as a process compressor or building, to a project involving a collection of assets distributed about a plant. Establishing the scope will determine how many hazard scenarios and potential impact outcomes will have to be considered.

4.3.1 Determine Hazard Scenarios

A hazard scenario is the chain of events starting with an initiating event typically, a failure in the case of an aging asset, that results in an undesirable consequence. Some examples of serious undesirable consequences include:

- Flammable or toxic vapor release
- Loss of primary containment due to fixed equipment failure (vessels, piping)
- Fired equipment explosion
- Significant process unit and/or plant fire or explosion
- Internal deflagrations and energy release
- Rotating equipment failures and energy release
- Electrical distribution system failures, including switchgear and transformer failure and explosions Loss of cooling water
- Structural collapse
- Loss of emergency shutdown capability
- Loss of fire and gas detection
- Loss of fire water capacity

As the examples show, a hazard scenario often involves the loss of process or structural containment or loss of functionality of a critical safety system.

The outcome of a hazard scenario consequence depends of additional consequential factors, such as ignition probability in the case of a flammable vapor release, or time at risk for non-continuous operations. For flammable vapor releases with immediate ignition, the outcome is a jet or pool fire with a local thermal radiation hazard. If the ignition is delayed, potential outcomes can include a vapor flash fire and burn hazard, or a confined or unconfined vapor cloud explosion with structural damage and life threatening hazards. What is important in defining the hazard scenarios is to ensure that all outcomes are properly identified and their significant consequences defined, with the aim to properly evaluate the risk.

4.3.2 Assess Consequences

This step involves assigning a consequence severity value to the scenarios. The level of sophistication available for determining hazard consequences ranges from qualitative judgement to quantitative computer models for fire, explosion and toxicity exposure (CCPS, 1999). Unless the risk impacts to potential resources being dedicated to the endeavor is quite large, the use of qualitative consequence criteria should be adequate. One place to find such criteria are risk ranking matrices similar to the one shown in Figure 4.1-2. Many companies have developed a standardized risk matrix that is used in conducting a process hazard evaluation, which defines hazard consequence categories. Other helpful resources for assigning consequence severity include:

- Plant incident reports and investigations
- Industry incident histories
- Case studies provided in this and other CCPS books

Once the scenario consequence values have been assigned, the next step is to decide whether a risk based decision is needed. This can be accomplished by comparing the assigned consequence to a set of criteria such as those shown in Table 4.2-1. If it is determined that RBD is not required, it may still be necessary to monitor the condition of the asset with the intent to address the problem at a later date. Alternatively, the magnitude of the fix may be sufficiently small to schedule repairs from existing or future maintenance budgets. Having followed a systematic approach this far means the issue is on the radar screen and a management strategy can be formulated.

4.3.3 Assess Likelihood

When the scenario consequence severity meets or exceeds the RBD criteria, an estimate of the likelihood or frequency of occurrence is required. As is the case with consequence, there are a variety of tools, both qualitative and quantitative, for estimating likelihood. As mentioned above, a properly constructed risk matrix is a suitable qualitative tool for risk ranking.

Another semi-quantitative technique is Layer of Protection Analysis (LOPA), which involves a more rigorous evaluation utilizing impact criteria (see above) and standardized failure rate data for initiating events and Independent Protection Layers (IPLs) (CCPS, 2001). LOPA is commonly applied for determining the Safety Integrity Level (SIL) requirement for Safety Instrumented Systems (SIS) per IEC 61511. (IEC 61511-1, 2016).

The American Petroleum Institute (API) Recommended Practice (RP) 580, *Risk-Based Inspection* (API, 2009) and 581 *Risk Based Inspection Technology* (API 2008) also provide guidelines for assessment of the risk of physical failure of equipment and piping. The Recommended Practices provide methods for determining the probability of failure as well as consequence assessment.

Other well established techniques include Failure Modes and Effects Consequence Analysis (FMECA) and Failure Modes and Effects Diagnostic Analysis (FMEDA). These techniques can be used to focus on specific failure modes and assess the consequences and frequency of occurrence. The latter is more often utilized to establish Failure in Time (FIT) for components (sensors, logic solvers and final elements) comprising safety instrumented systems. Other applicable references for additional sources of information on these and the other techniques discussed are, British Standard 7910 *"Guide on methods for assessing the acceptability of flaws in metallic structures"* (BS, 2013), API RP 579-1/ASME FFS-1, *"Fitness-for-Service"* (API RP 579, 2007) and *"Prioritization of Safety Related Plant Modifications Using Cost-Risk Benefit Analysis"* (Stephens, 1992). Some of these techniques are much more rigorous and require significant resources to apply.

4.3.4 Determine Risk

Using the established scenario consequence severity (Step 2) and likelihood (Step 3), the risk level can be found using the risk matrix. The tolerable risk target is set to achieve ALARP or placement in the green tolerable region of the risk matrix. Mitigating the risk to the tolerable range can be achieved either by lowering the frequency or reducing the severity of the consequence. In the case of catastrophic failures, the consequences are not easily mitigated, hence the need for multiple independent layers of protection. However, in the case of an aging asset, replacement of the asset may result in some combination of consequence reduction and resetting the clock on frequency.

Using the scenario consequence severity and frequency estimates in combination with the risk raking matrix will determine whether a gap exists between the scenario risk and the tolerable risk region. In general, the frequency gap between where the scenario risk lands on the matrix and the tolerable region needs to be managed by applying additional safeguards and IPLs, or modifying the risk by proper repair or replacement of the asset.

Risk assessment does not make the problem disappear. It merely provides a clearer vision of what could happen so that better decisions can be made. A knowledge of risk helps clear the fog and helps in making sound decisions based on a systematic approach and data.

4.3.5 Develop Risk Mitigation Controls

Once it has been determined that the risk needs to be managed, the focus turns to considering effective options and the demand on company resources. This process needs to start with mustering the appropriate technical resources. This should start with the people who have been managing the Asset Integrity program. They would have the most intimate knowledge of the current condition of the asset of interest, and possess the technical skills to undertake further evaluations, like FFS evaluation, and possible repair options. Company knowledgeable experts or third-party subject matter expert are another resource that can be enlisted to address complicated evaluations and complex repairs.

One of the key questions that may arise at this or an earlier stage is whether the asset is fit for purpose. There are several peer reviewed methodologies both domestic and international for evaluating pressure containing equipment to determine fitness for service. One of the most commonly used is API RP-579, *"Fitness-for-Service"* (API, 2007)/ASME FFS-1 Another is BS 7910 published by British Standards for application to metallic structures across a range of industries and is therefore more general in its approach than API 579 in assessing the acceptability of flaws in metallic structures. For example, the procedures for assessment of fracture, fatigue, flaws under creep conditions and other modes of failure, may be equally applicable to non-pressure containing equipment.

API 579 has modular organization based around each defect/damage type. Each module generally has three levels of assessment, with each level involving an increased level of analysis sophistication.

- Level 1 is aimed at inspectors for use on site for quick decisions with the minimum of data and calculation
- Level 2 is intended for skilled engineers or other qualified technical personnel, and requires simple data and analysis
- Level 3 is an advanced assessment requiring detailed data, computer analysis and considerable technical knowledge and expertise in FFS assessment procedures

It is quite possible and appropriate that a Level 1 FFS be performed as a pre-screening exercise, to determine the severity of the defects before the RBD process is applied. Of course, this presumes that the asset was undergoing scheduled inspections with recorded asset integrity data. At this step in the RBD process, for potentially major Capital Expenditure (CAPEX) projects, a more detailed Level 2 or 3 FFS evaluation may be beneficial to determine with a greater degree of certainty, how significant the damage is. This can involve using some of the computational techniques presented in the RP or in some cases the application of computer models like finite element analysis.

Other factors that need to be considered in selection of the control option or options are lifecycle extension and cost. This attempts to address questions like:

- If we repair, how long is the equipment life extended based on the remaining life assessment?
- Should we replace it now when management is focused on the issues instead of waiting and risking additional delays and safety?
- How much do we save by repairing vs replacing?
- Should we run it to the end of its lifecycle and then replace? (How sure are we that we know when the safety limit is reached?)
- Can we afford to incur a total system outage or failure?

A key theme in this regard is the expected life extension of the asset. Like buying a new or used automobile, it's all about how much mileage you expect to get for the money spent.

Deliberately running equipment to its safe operating limit, can be a strategy for some equipment. An example might be rotating equipment with installed redundancy. In following this strategy, it is necessary to have a comprehensive asset integrity program for the asset that can predict with a high degree of certainty that the safe operating limit has not been exceeded. For some damage mechanisms, such as CUI or cyclical metal fatigue, the degree of certainty required may not be attainable.

4.3.6 Implement Risk Controls

Once the risk controls have been developed, a plan and schedule for implementation should be established and acted on. Ultimate success depends on willingness to act and take corrective actions. Problems with implementation are likely to arise when the future of the operation is uncertain due to unwillingness to spend resources fixing equipment that may to sold off or shut down. However, transferring aging equipment problems to the next owner may not absolve the company from liability depending on how the transfer agreement is written and how thorough the due diligence has been performed.

4.3.7 Information Required for Risk Based Decisions

Risk based decisions rely on factual data and interpretation. Factual data encompasses all that is known about the condition of the asset up to the decision time. This would include:

- Initial design and construction information such as specification data sheets, U-1 forms, shop testing and inspection records
- Service and operating history including changes in service conditions impacting the integrity operating window and service aging
- In-service inspection and testing records, NDT, TML, visual, vibration, etc.
- Repair/replacement history

- Incident history, unexpected failures, unscheduled outages
- Management of change documentation

In addition, outputs of Process Hazard Analysis (PHA), Process Safety Management (PSM), jurisdictional requirements and Climate/Environment Changes can be considered to provide input to programs such as Risk Based Inspection (RBI).

Proper interpretation of the data is essential to attain the correct conclusions. If the problem is of a nature that the technical resources at the facility are challenged, subject matter experts from other company facilities or external specialists should be consulted to incorporate a broader range of experience into the evaluation.

4.3.8   Documentation of Risk Based Decisions

Risk based decision making follows a systematic approach with inputs from a variety of sources and sub activities, and warrants suitable documentation. The documentation can more or less follow the RBD steps in Figure 4.3-1, beginning with the members of the team who participated in the process. Table 4.3-1 provides some guidance on how to document the RBD.

Table 4.3-1. RBD Documentation Guidelines

Step	Activity	Appropriate Documentation
1	Define Scope and Scenarios	Describe the assets that are the subject of the RBD, the failure mechanisms of concern, and what would happen if the asset failed or lost functionality. For scopes with multiple equipment items, each unique asset should be addressed individually.
2	Assess Consequences	Document the consequence categories considered for each scenario and the severity rating assigned. Describe any hazard quantification tools/models employed or references utilized to characterize the consequences and key assumptions made.
3	Assess Likelihood	Assign frequencies to the various consequences outcomes and explain the methodology(s) used to arrive at the values.
4	Determine Risk	Reference the company's risk management criteria, and record the tolerable frequency target value for each consequence. Record the frequency assigned to the scenarios, compare it to the tolerability target, and identify which scenarios require additional control measures.
5	Risk Control Options	Describe what control options were considered and rejected, which ones were selected, and what factors were considered in arriving at the final action. Explain how life extension and cost estimates were developed.
6	Implement Controls	Prepare an implementation plan with defined actions and completion dates, and track to closure.

4.4 EMBRACING RISK BASED MANAGEMENT

The purpose of risk based management practices is to focus an enterprise's resources in areas that have potential to obtain the most benefit, whether that be increasing profitability, improving safety, or reducing inefficiency. To be effective, the concept needs to permeate the organization from upper management to those managers and supervisors that implement the action required to reap the benefits. Managing the risks of aging assets is not that different, except there tends to be a mentality of "If it isn't broken why fix it?" That is precisely the question that RBD can help answer. Few companies can afford to fix everything that is in need. RBD properly applied, will allocate resources to the right places at the right time.

4.4.1 Alignment of Management and Operations with Risk Based Decisions

Some of the Risk Based programs for asset integrity management of process equipment are implemented at a level in the organization below upper management namely Risk Based Inspection. Managers with budget approval authority may not be as familiar with these techniques, although the objective of effective allocation of resources is the same. When it comes to aging assets, an important element is communicating the risk convincingly to management.

A well developed and documented RBD process can be a good risk communicating mechanism to management. The process is systematic and reasonably rigorous, and if properly presented, the message should resonate with management. It will show by concrete example how some of the risk management methods and tools are applied and the clarity they bring to the decision process. It may also be informative to management that while some recommended actions need to be funded soon based on the findings, others posing lesser risk can be deferred to a later time.

4.4.2 Incorporate Corporate Responsibility and Economic Value

Proactively managing an effective process safety program displays a high level of corporate responsibility and encourages individuals to sustain it long-term. Thus, a robust process safety program will help your company reduce risk and avoid loss by providing enhanced risk reduction:

- Lives are saved and injuries are reduced
- Property damage costs are reduced
- Business interruptions costs are reduced
- Loss of market share is reduced
- Litigation costs are reduced
- Incident investigation costs are reduced
- Regulatory penalties are reduced
- Regulatory attention is reduced

Implementing an effective process safety program which includes a viable asset integrity management program, helps to create and sustain value for the company and its shareholders. Additionally, embracing process safety as an essential part of the company culture, allows companies to achieve a measurable increase in revenues and a reduction in costs. The result is improved asset integrity, reliability and resilient operations. The following are returns from the investment in process safety (CCPS, 2006):

- Productivity Increases
- Production Costs Decrease
- Maintenance Costs Decrease
- Lower Capital Budget Required
- Lower Insurance Premiums

4.5   DEALING WITH UNEXPECTED EVENTS

With the increasing use of Probability Centered Maintenance and Risk Based Inspection, there is the possibility that an aging asset was improperly classified as lower risk and did not receive the attention it should have. This situation could result in an unanticipated failure event.

What steps can be taken to minimize surprises? These practices involve making informed judgement regarding risk parameters (failure severity and likelihood). Certain assets that were initially characterized as having lower risk should be periodically re-assessed to make sure the severity and probabilistic assumptions are still valid. One trigger for the re-assessment is the next scheduled inspection involving measurement data on failure mechanisms. This allows incorporation of the latest factual data into the risk assessment process. Most Maintenance Management System (MMS) software includes capability for trending corrosion rates and projecting remaining life. For other failure mechanisms (e.g., fatigue), a more experience based assessment may be required. In some cases, proscriptive time interval, such as 5 or 10 years would be a better choice.

Sometimes the surprises happen in orphaned assets like structures, and ancillary support systems including drains, underground piping, etc. Sometimes what appears as a minor failure in such a system can result in a serious incident.

A case in point involved a cast iron drain line at an elevated plant parking lot, in a northern location where de-icing chemicals were used. After some years, an elbow in the drain line corroded out from attack by chlorides in the de-icing salts. There was a pipe way next to and below the level of the parking lot, containing a high pressure insulated hydrogen line. The piping orientation was such that water from the failed drain line above dripped onto the hydrogen piping and eventually caused CUI and weakened the pipe. The pipe eventually failed catastrophically resulting in an explosion and fire. There was a warning sign in the form of staining on the aluminum cladding on the piping, which was included in an asset integrity inspection program. But the connection between the leaking drain and CUI was never made and the drain was never repaired. There can be some lessons to take away from this incident.

- Fully understand the process hazards. Do not underestimate the hazard potential of any asset or equipment before it is properly assessed. Failure of structures and non-process supporting equipment may impact the known hazardous equipment in subtle ways.

- Be on the lookout for equipment conditions that appear abnormal or have changed from normal operating conditions. The staining on the insulation cladding was a warning sign of possible CUI.
- Personnel associated with operating and maintaining the plant assets should be engaged in some form of troubleshooting and reporting of abnormal or unusual conditions. However, to be effective, the personnel not directly involved in maintenance and inspection need to receive instruction in basic failure modes and warning signs.

4.6   RISK BASED DECISIONS SUCCESS METRICS

The success of Risk Based practices will be measured in terms of money spent and benefits obtained. Money spent would include annual expenditures for

- Probability Centered Maintenance (PCM)
- Risk Based Inspection
- RBD Scheduled repair/replacements

Benefits over time that can be measured and tracked would include avoidance of:

- Incidents due to failure of worn out aged assets
- Direct cost of incidents (breakdown repairs)
- Indirect cost of incidents (business interruption, increased insurance premiums, cleanup costs, fines, etc.)

Of course, these represent lagging indicators, which would hopefully be declining over time. Money spent to avoid future costs is harder to quantify. Avoidance cost is a concept used in the insurance industry to set premiums. It takes into account the value of possible future losses, the probability of such losses and utilizes a Monte Carlo simulation to determine the probable average annual loss potential. If a company has a portfolio of potential hazard scenarios with associated consequential financial losses and an estimate of the probabilities for those losses, the same methodology can be applied. There is a positive benefit when the cost estimate for addressing the consequences is less that annual probable avoidance cost (Stephens et. al., 1992). Additional examples of metrics are illustrated in Table 4.6-1.

Table 4.6-1. Corrective and Preventive Metrics Definitions

Metric	Definition
Reactive Maintenance (RM) backlog	Number of open RM's for each unit in the facility
RM aging	Average age of open RM's for each unit in the facility
# of RM's open for corrosion related issues	Number of open RM's, for each unit in the facility, related to cathodic protection or corrosion related issues
System Preventive Maintenance (PM) schedule compliance for corrosion related issues	% of PM's performed on-time for cathodic protection or corrosion related issues

# of overpressure events	Number of overpressure events for each unit in the facility
# of open Process Safety (PS) recommendations	Number of open Process Safety recommendations, from PS Audits, requiring action by each unit in the facility

Table 4.6-1. Corrective and Preventive Metrics Definitions, continued

Metric	Definition
# of system shutdowns	Number of system shutdowns as measured by count of trips of running units, Emergency Shutdowns (ESD) in standby, and failed starts
# of system defined obsolete component types	Number of identified obsolete component types (by make and model) included in the obsolete equipment list (not a complete count of individual equipment items)
Compressor efficiency	Efficiency ratio. Calculation based on inlet and outlet temperatures and pressures
Engine fuel efficiency	Ratio of fuel cost to work Horse Power (HP) hrs

5

MANAGING PROCESS EQUIPMENT AND INFRASTRUCTURE LIFECYCLE

5.1   LIFECYCLE STAGES

Processes go through various stages of evolution research, process development, design and construction/fabrication, commissioning/startup, operation, maintenance, and finally decommissioning. Progress through these stages is typically referred to as the process lifecycle. The CCPS concept book *"Inherently Safer Chemical Processes"*, devotes a chapter to the topic of lifecycle stages (CCPS, 2008). The goal is to continuously apply the strategies and practices of safe facility operation and maintenance, throughout the lifecycle stages, to guarantee the protection of employees, the public, and the environment. With proper management, aged assets can be safely operated in a culture that demands safe and reliable operation.

5.2   ASSET LIFECYCLE MANAGEMENT

Asset Lifecycle Management (ALCM) is a comprehensive approach to optimizing the lifecycle of assets beginning at conceptual design, continuing through construction/fabrication, commissioning/startup, operation, shut down and decommissioning. Thorough planning, analysis and timely execution allow appropriate data-driven decision-making to occur and enable ALCM to achieve optimum:

- Operating and maintenance strategies
- Organizational structure
- Staffing and training requirements
- Optimized asset integrity practices

ALCM involves a holistic approach to achieve effective asset investment decision-making that addresses not only process and infrastructure assets, but also the supporting resources, business processes, data and enabling technologies that are critical to sustainable success.

This all-encompassing approach to asset lifecycle management enables vast amounts of asset data, particularly from the asset integrity management program to be effectively managed and leveraged at a practical ongoing business level. Figure 5.2-1 shows an example of asset lifecycle.

While an asset spends the majority of its life managed in the operate/maintain phase, its integrity management has to begin with its conception. How well it performs depends in large part on the Process Technology (PT), how consistently the process is operated and the Quality Assurance (QA), during design and construction phases. Technology determines what the equipment is to be (specifications, acceptable limits). QA ensures that it starts out that way, then proper operating procedures and training make sure it is operated within acceptable limits. Finally, a good asset integrity program maintains it within those limits. Figure 5.2-1 depicts how all these facets of asset integrity management can interact over the entire asset lifecycle.

Some of the key elements that support lifecycle asset management include: management strategy, optimum organizational design and long term asset planning.

5.2.1 Management Strategy Development

The management strategy should encompass a shared vision, strategy and action plan for a successful asset lifecycle management program. Developing a vision brings organization stakeholders together to create a common understanding of asset management, reach consensus on enterprise objectives and prepare a plan for program implementation. The outcome of this process should be an asset management strategic plan that defines an action plan, lays out an implementation schedule, addresses budget requirements, and states the business case for moving forward with a viable asset lifecycle management process.

5.2.2 Organizational Design

To achieve ALCM objectives, the organizational structure should enable the right people, processes, data, and information technology, to come together at the appropriate time. This in turn requires that organizational roles and responsibilities are defined in a structure that permits qualified resources to be available to achieve program objectives.

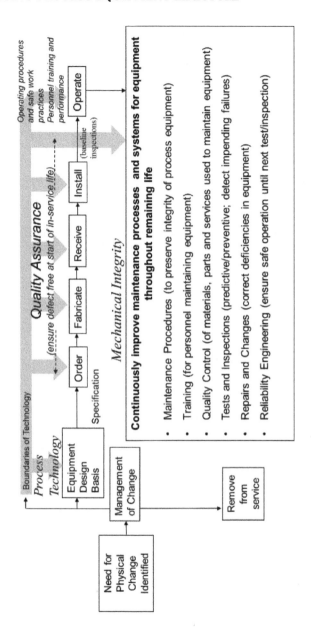

Figure 5.2-1. Asset Lifecycle Management

5.2.3 Long-Term Asset Planning

Deciding how to best invest limited capital, and operations and maintenance funding is central to managing the lifecycle of assets.

Understanding the current condition and capacity of the facility's assets, as well as future capacity and reliability requirements is essential. Also required is an understanding of the cost and risk associated with implementing or deferring asset repairs or replacements as seen in Chapter 4. The planning process should address:

- Prioritize asset capital projects in the short term (e.g., five-year period) based on strategic objectives
- Forecast capital renewal and replacement costs over a longer period (e.g., ten to fifteen years)
- Assess aging equipment and infrastructure funding requirements against long-term revenue and cost forecasts

5.3   GENERAL TOPICS

5.3.1 Manage by Operational Integrity

5.3.1.1 Safe Operating Limits

An essential aspect of lifecycle management is adhering to the safe operating limits as part of the integrity operating window for the asset. For process equipment, safe operating limits are addressed in the Process Knowledge and other elements of CCPS' Risk Based Process Safety Guidelines (CCPS 2007). The requirement is for information pertaining to the technology of the process including, safe upper and lower limits for such items as temperatures, pressures, flows or compositions, and, an evaluation of the consequences of deviations as part of Management of Change, including those affecting the safety and health of employees. CCPS defines safe operating limits as limits established for critical process parameters, such as temperature, pressure, level, flow, or concentration, based on a combination of equipment design limits and the dynamics of the process (CCPS, 2007). Another phrasing of the concept from an operating company states, safe operating limits are intended to define the ultimate safe operating conditions, based on the most constraining of either physical equipment limits or process limits (Conoco/Phillips, 2013). The intent is that a safe process operating envelope needs to be defined and documented taking into account information like Maximum Allowable Working Pressure (MAWP), Maximum Allowable Working Temperature (MAWT), conditions that result in loss of process control (e.g., approach to runaway onset, etcetera).

This information is then to be used in the development of operating procedures, which should address operating limits, consequences of deviation and steps required to correct or avoid a deviation. This safety requirement is included to ensure that operators are adequately trained and re-trained on standard and emergency operating procedures for normal and upset conditions. These procedures need to take into account the critical operating parameter limits and what actions should be taken in the event that these limits are approached or exceeded, to avoid exceeding the design limits of equipment.

For pressure equipment, API RP 584: *Integrity Operating Windows* (API 2014) addresses defining, monitoring, and responding to deviations from preset limits on operating variables established and implemented to prevent potential breaches of containment that might occur as a result of not controlling the process sufficiently to avoid unexpected or unplanned deterioration or damage to pressure equipment. Operating within the IOWs should result in predictable and reasonably low rates of degradation.

The CCPS RBPS (Risk Based Process Safety) elements are designed to avoid operating conditions that could imperil the short term or long term integrity of process equipment. Although structures and ancillary infrastructure are not explicitly covered by Process Safety Management PSM), the concepts are still examples of good Asset Integrity Management (AIM). This could involve identifying and documenting parameters such as weight limits, cycle limits, voltage/current limits, torque limits, etc. for infrastructure. Even for companies or facilities where PSM is not a compelling requirement, applying these tenants should be embraced as a means of managing the premature aging of assets.

5.3.1.2 Mechanical and Functional Integrity

Maintaining equipment and asset integrity throughout its lifecycle is a primary goal for avoiding incidents with significant consequences. Asset integrity is one on the cornerstones of common process safety regulations. It involves monitoring the health of an asset through collecting, recording, tracking and analyzing data on its physical and functional condition. Incorporated into the management of asset integrity are many codes and practices developed by professional societies and industrial associations, which cover various lifecycle stages including design and construction, maintenance and inspection, modification and decommissioning. This body of expertise is commonly referred to as Recognized and Generally Accepted Good Engineering Practices (RAGAGEP). It forms the basis for programs developed to address maintaining the physical and functional integrity of process assets. RAGAGEP has also been incorporated into government regulations by reference. The graphic shown in Figure 5.3-1 summarizes many codes and standards applied to facility assets at different lifecycle stages that are considered RAGAGEP. Chapters 6 and 7 provide more specific information on RAGAGEP of particular relevance to managing integrity of aging assets.

Situations exist in older facilities where equipment that does not meet current codes is grandfathered for continued use through a special provision. It is important that these provisions are documented and included in the Process Safety Information (PSI), and that the compliance with the provisions is being achieved.

5.3.2 Managing Change During Lifecycle

The asset integrity of equipment that experiences an extended lifecycle could undergo a number of changes. Some of these changes are consequential - simply the result of continued long term exposure to steady state and changing process conditions. Other changes are human initiated and are driven by the need to address problems as they arise in an operation. Some of these are maintenance related while others involve engineering.

Having a management system to review and track these changes and manage deficiencies is necessary to ensure that they are not compromising the

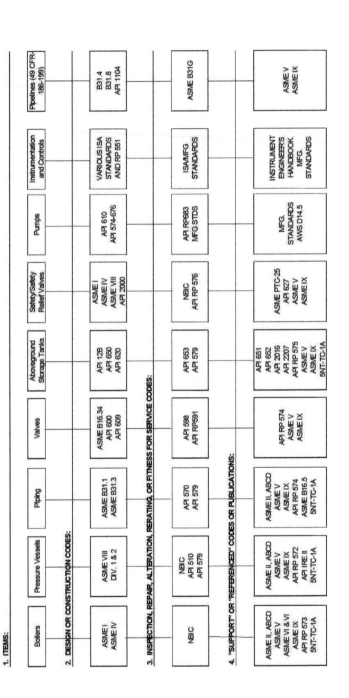

Figure 5.3-1. Codes and Standards Applied to Facility Assets

Source: Guidelines for Asset Integrity Systems –CCPS Publication (CCPS, 2006)

integrity of a system or piece of equipment. Typically, MOC procedures address component changes not in-kind, or larger projects covering new equipment installation. Changes in process technology can result from changes in production rates, raw materials, experimentation, equipment unavailability, new equipment, new product development, change in catalyst and changes in operating conditions impacting the safe operating limits and integrity operating windows, to improve yield or quality. Equipment changes include, among piping pre-arrangements, experimental equipment, computer program revisions and changes in alarms and interlocks. Additionally, there are other areas in which a management of change is also needed:

- Changes in procedures (e.g., standard and emergency operating procedures, safe work practices or maintenance and inspection/test procedures)
- Changes in Inspection, testing, preventive maintenance and repair requirements (e.g., changing inspection interval, or the lubricant of a compressor or pump)
- Changes in site infrastructure, such as electrical distribution systems, support systems, fire protection or fixed and/or portable buildings
- Management of Organizational Change

Therefore, some process safety regulations require establishment of means and methods to identify and manage changes through a proper MOC procedure, including the option to conduct a process hazard evaluation of the change to address its impact on safety and health.

Changes are sometimes executed in parallel with one another by different parties to solve specific equipment problems. For example, one MOC is addressing a control system issue, and other is dealing with a relief system capacity problem. Consequently, the solution to one problem may contribute to another. This, in part, explains the importance of conducting periodic hazard evaluations on an entire system to provide a holistic perspective. After many MOCs have been implemented in a specific area, it may be necessary to completely redo a hazard evaluation to effectively address the risks of all the cumulative changes. Some of the changes to aging equipment are more subtle and longer term. How well do the process safety management system procedures handle such changes? Are there procedures to periodically check equipment service conditions against its original design intent?

For some aging mechanisms like metal loss due to corrosion/erosion, it is common practice to take Thickness Measurements (TM) and compare them to minimum wall thickness values established from design calculations and fabrication drawings, or standard thickness specifications (e.g., piping schedules). Generally, management of change does not address ongoing lifecycle changes in assets on a periodic, holistic basis. For example, during an extended lifecycle of a facility or piece of equipment there will likely be many changes, involving both physical and service conditions. Have such changes been tracked and has the cumulative effect of these changes been considered and analyzed? The MOC process hazard analysis used to evaluate a change prior to its execution will not typically address this aspect. As the facility or asset approaches its intended service life, triggered by time, repair history and the number of MOC modifications, a more global MOC to consider the cumulative effect of the changes would be appropriate. The MOC procedure should be aligned with such a need.

In some cases, temporary repairs (e.g., adding bypass piping or installing a pipe or flange clamp to alleviate an upset) are left in place permanently and their removal is overlooked. With time these changes are often forgotten, especially if they are never added to the P&ID, and are not considered during the hazard identification revalidation. Unless the asset integrity program has a system to verify that deficiencies are properly managed, tracked and trended, and that temporary and permanent repairs are evaluated as part of the program, equipment reliability can suffer as a result. This can increase the potential for an equipment breakdown. Changes that impact the asset integrity program should be considered as part of MOC The MOC system is the usual way of ensuring that temporary repairs are addressed in a timely manner. When a temporary or emergency repair is required, a MOC request should be submitted with a date for removal of the temporary change or converting the change status to permanent. If it is made permanent, the PSI should be updated to reflect the change before the MOC is closed.

5.3.3   Orphaned Assets

A fundamental cause or contributor to some of the aging asset problems encountered, is lack of ownership. This is more often the case with ancillary systems or infrastructures which are assigned to utility or logistics oversight or not covered at all (sometimes referred to as grey zone equipment). Examples of grey zone equipment include pump ancillary piping, process hoses, interconnecting (OSBL) piping and supports, level bridles and associated piping, piping and supports to/from remote storage, vendor operated equipment (e.g., treatment chemical equipment), and pipelines supplying the plant. The assets may not be in the equipment inventory of the formal asset integrity program. Lack of ownership leads to neglect. Operations and Maintenance should periodically compare the enterprises asset inventory list with the asset list in the asset integrity program to ensure all assets meeting the established screening criteria are accounted for.

5.3.4   Disrepair of Assets

Disrepair connotes equipment that has been neglected, poorly or inadequately maintained. When inadequate or improper repairs are made, equipment degradation can continue unnoticed and a significant failure may occur. Equipment in a state of disrepair is more likely to occur towards the end of its lifecycle.

Some ways to prevent and address disrepair due to neglect are mentioned above. Make sure that asset ownership is clearly defined, and that an asset integrity program is followed. Inadequately repaired equipment is more of a human resource quality issue or possible insufficient maintenance budget. Regarding the latter, if adequate funding for proper maintenance of higher risk assets cannot be allocated, then management should be informed of the risk of continued operation.

Improper maintenance practices may also occur due to lack of knowledge and skills of personnel assigned responsibility, and implementation of maintenance and testing. Along with assigning roles and responsibilities, management's responsibility is to ensure the persons filling positions have the necessary background, skills and certifications to perform proficiently. As the example in Chapter 3 illustrates, assigning inexperienced recent graduates to the inspection department for on-the-job training, without proper training and certification, can lead to substandard performance and potential danger. Training requirements for managers, supervisors, inspectors and technicians engaged in managing the facility assets should be defined. The use of training or competency matrices, can assist in organizing and ensuring personnel have or are receiving all the training required for the position.

Can proper repairs and maintenance overcome years of neglect? There is no single answer to this however, an approach to help arriving at an answer is possible. The first step is to assess the current condition of the disrepaired asset. This may take some effort depending on the quality of record keeping, last known inspection, current service, etc. Some form of inspection will almost certainly be required. Depending on the type of asset (e.g., process equipment, infrastructure, building) this may involve, external or internal visual inspection, external or internal nondestructive testing, inspection under insulation, remote internal visual inspection (e.g., borescope) for example. If baseline testing results are not available, referring to original design specifications and drawings will be necessary.

The next activity is the review of the inspection and testing reports and evaluation of the results. The depth of the evaluation will depend on the seriousness of the disrepair. The reports themselves may include recommendations such as rust and scale removal and recoating for limiting degradation. In moderate to severe cases, some additional data collection and engineering calculations will be required (e.g., remaining thickness, API 579-1/ASME FFS-1 level 1, 2, or 3). These evaluations should produce a plan for addressing the service ability of the asset, which should include the required refurbishment of the equipment, and the maintenance and inspection regime going forward.

Last but not least is estimating the lifecycle extension and cost for restoring the health of the asset. In many cases, the cost/benefit ratio will be in favor of refurbishing and continue to maintain. For more significant repairs (weld overlays, shell plates, replacing major equipment support structures in an operating plant), where the lifecycle extension is less certain and the cost is high, total replacement may be the cost-effective option.

In the case of poorly maintained control systems, the decision may be driven by the age and functionality of the equipment. Process control instrumentation has undergone significant technological advancement in recent years and that threat will likely continue. Some suppliers no longer support some lines of older products. For this type of asset, replacement may be the only realistic option to avoid continued disrepair as described in Section 5.3.6 about Cannibalism.

5.3.5 Extending Lifecycle with Rebuilt Equipment

When aging equipment is rebuilt does it really start a new lifecycle? To answer this question, it's useful to consider the failure in time distribution function which is:

$$F(t) = 1 - \exp(-\lambda t)$$

where,

λ is the failure rate

t is time

The distribution for one device is shown in Figure 5-3.2. The example given is for proof testing of a safety instrumented system, however, the same concept is mostly applicable to an aging asset.

The curves are idealistic in that the underlying assumption is that the proof test coverage is 100%. In reality, there can be residual fault mechanisms that are not detected by the proof test. Hence the baseline probability is not returned to

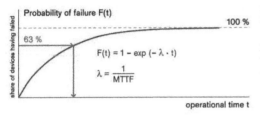

Probabilities to fail.
The probability of failure of one component or device with the failure rate λ continuously rises during operational time. At the time of the MTTF the probability of failure is 63 %.

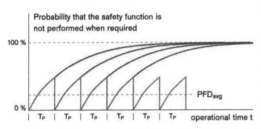

The probability of failure that in case of demand the safety function cannot be performed also rises continuously. If however the successful perform-ance of the safety function is demon-strated by regular proof tests, then at the time of test the probability that the system will perform correctly is 100 %, meaning that the probability of failure has been reset to zero with each successful proof test (blue zigzag-curve).

Note: Minimum Time to Failure (MTTF), Probability of Failure on Demand (PFD)

Figure 5.3-2. Probability of Failure vs. Time for a Safety Instrumented System (Dräger, 2007)

zero, and increases exponentially with time along a similar path (CCPS 2016). Similarly, for a rebuilt asset, one would not expect that all the possible fault mechanisms accumulated prior to rebuilding are eliminated. This is not to say that rebuilding an asset is unsafe or undesirable only, that it may not be returned to its condition when new.

That suggests that any integrity monitoring program in place prior to rebuilding should be continued.

Refurbished assets, which were less extensively repaired, may have similar issues. Determining whether they are fit for service and setting service limitations can be required. Fit for service evaluation is possible for steel structure and pipe bridges using standard stress analysis techniques like finite element analysis. However, for some infrastructure assets, there are no proscribed standards for FFS evaluation.

5.3.6   Managing Used or Refurbished Equipment

Cannibalism occurs when spare parts are not readily available and parts are sometimes removed from companion equipment. Sometimes companion equipment has already been abandoned and sometimes it is a spare unit. Cannibalism is seldom practiced with new equipment. With aged equipment, this practice is more common and re-used (shuffled) parts may not fit the equipment or function as well as the originally specified parts. This practice is often done with limited control over the quality of the substituted part.

Another practice is to purchase rebuilt or "re-furbished" parts from used equipment suppliers. The risk in these practices is the introduction of a component that does not materially extend the system lifecycle, and may actually reduce it. They are often stop-gap fixes to get production back online.

The aim is not to prohibit the use of used or refurbished equipment, but rather to control the risk. This can be accomplished by applying some management controls on the acceptance (qualification) and inventory management of such equipment. The first part means having a procedure and criteria for determining the suitability and serviceability of the items. For example, does the item meet the specification requirements of the equipment it is replacing? Next, is the condition of the item fit for the service? In the case of rebuilt equipment, what quality checks has the supplier performed to warrant the serviceability? For cannibalized parts, what precautions have been taken to determine fit for service?

In the case of planned use of a used equipment item, there are other PSM systems that should apply, namely MOC and Process Hazard Analysis (PHA). These elements allow a team with a diversity of skills to evaluate the condition and risk of installing equipment that is not new.

After acceptance of the used and refurbished parts, there needs to be a system to manage the qualified inventory and properly store it. This would apply to items that are not acquired for immediate installation. When cannibalism is anticipated, (e.g., control instruments no longer supported), the components should be subjected to the acceptance process and then added to the acceptable inventory, placed in appropriate packaging if required, and stored away from the elements.

5.3.7   Mothballing and Re-commissioning of Aged Assets

It is necessary to maintain the asset integrity of assets, even when decommissioned and mothballed. Equipment should be de-energized, de-inventoried, cleaned and additional measures should be taken for equipment preservation and any ongoing inspections (CCPS, 2006). If the intent is to return the asset to service later, the mothballing should include purging and other measures to help preserve equipment such as maintaining a proper atmosphere to prevent corrosion (CCPS, 2006).

Re-commissioning of mothballed assets is a change in service (from idle to productive service) and should be managed with an MOC system. In addition, a re-commissioning procedure should be followed for such an occurrence. The procedure may also include a change-of-service (e.g., new chemistry, etc.) approval that should consider:

- Length to time out-of- service
- The extent to which ongoing inspection and/or PM was performed
- New process conditions
- Re-rating

It should also address inspection requirements and other equipment checks (e.g., instrumentation, relief capacity, piping stress) to verify that the used equipment is suitable for its intended use (CCPS, 2006). RBPS Operational Readiness element requires a safety review prior to startup (CCPS 2007).

5.3.8 Partial Upgrades to Older Facilities and Equipment

Industry codes and practices continue to evolve while incorporating positive and negative learnings from industrial history. Some new and revised codes include safety features aimed at preventing incidents that have resulted in serious losses. There is a provision in some codes that permits "grandfathering" of pre-existing facilities and equipment this allows older equipment to remain "as is" provided it can be operated safely. The decision to grandfather any facility or piece of equipment should be made through consultation with regulatory authorities.

One equipment category which has progressed significantly in recent years, is that of the fired heaters. These are a major hazard source on many plant sites. Today's process heaters are often larger and more complex than the previous counterparts. Burner management systems are now designed to ensure that safe conditions exist at all points within a heater. API RP 556, *Instrumentation and Controls for Fired Heaters and Steam Generators,* provides detailed guidelines for designing and operating the control systems on modern heaters. This is only one example of recent changes made to this category of equipment.

While there is no imperative to adopt these modern features on older heaters, some companies have done partial retrofits to existing equipment. This may be done with good intentions however, if partial upgrades are not done consistently across a site a wide range of equipment generations may exist constituting a safety hazard (human error) to operating personnel. If a company or facility wishes to partially adopt new codes and standards on a voluntary basis to grandfathered facilities and equipment it should set some strict uniform standards and adhere to these. The rationale for making such upgrades should be handled by MOC and formally documented in company archives.

5.4   PREDICTING ASSET SERVICE LIFE

5.4.1 Mean Life and Age

One method of estimating service life is from equipment mortality data for a population of similar equipment. The mean life expectancy can be obtained by summing the lifetimes and dividing by the number of casualties.

"The essential weakness of the sample mean is that it only uses information of components that have died. For an equipment group with very few dead members, surviving components, as well as dead, contribute to the mean life. The approaches that are based on the Weibull or normal probability distribution have been developed to estimate the mean life and its standard deviation. The merit of the probability-distribution-based approaches is due to contributions of both dead and surviving components to the mean life being taken into consideration. Even with limited data of components that have died, the models can also produce a relatively accurate estimate" (IEEE, 2006).

The limitation of using mean life as a measure is illustrated in Table 5.4-1 for a group of 100 500-kV reactors at the British Columbia Transmission Corporation

(BCTC). In this example, there are only four retired samples in the total of 100 reactors having a calculated mean life of 25 years.

However, 35 reactors of the group already exceeded 30 years confirming that the estimated mean life of 37 or 38 years using the distributions should be much more reasonable than the 25 years (IEEE, 2006).

Table 5.4-1. Estimated Mean Life for the 500-kV Reactors (IEEE, 2006)

	Normal	Weibull	Sample Mean
Mean Life (years)	37.628	38.363	25.0
Standard Deviation (years)	6.896	6.293	-

There are two concepts related to measuring the age of infrastructure assets: natural age and functional age. The natural age is the difference between the in-service date and the present date, which is easy to calculate.

For the purpose of system planning, a rough estimate is generally sufficient and the natural age can be used. In maintenance, however, the focus is typically on a specific piece of equipment. In this case, it is better to obtain an estimate of the functional age, which depends on the deterioration status associated with usage history and operating and environmental conditions. The functional age can be estimated through a field assessment in some cases (IEEE, 2006), including monitoring through an inspection and testing program.

5.4.2   Assessing End-of-Life Failure Probability

With the estimated mean life and age of a specific piece of equipment, its aging status can be qualitatively judged since we know how far away it is from the mean life. The reason for concern about the aging status is the potential risk associated with end-of-life failure of aged equipment. In order to quantify the risk of aging failures, it is necessary to assess the end-of-life failure probability of aged equipment.

As is well known, the relationship between the failure rate or failure probability and the age can be graphically expressed using a so-called basin or bathtub curve, as shown in Figure 5.4-1. It can be seen from the figure that the failure rate at the wear out stage increases dramatically with the age. In fact, the bathtub curve can be mathematically modeled using a Weibull or normal distribution. Figure 5.4-2 shows the relationship between the failure rate and age for a normal distribution failure density function, Figure 5.4-3 provides the same relationship for a Weibull distribution failure density function. The μ and σ in Figure 5.4-2 are the mean and standard deviation of the normal distribution, whereas β and α in Figure 5.4-3 are the shape and scale parameters of the Weibull distribution. It can be seen that the relationship shown in the two figures is consistent with that expressed in the wear-out stage of the life bathtub curve. Note that the Weibull distribution can be used to model all the three portions of the bathtub curve: $\beta < 1$ for the infancy stage, $\beta = 1$ for the normal operating stage, and $\beta > 1$ for the wear-out stage.

There are two failure probability concepts for end-of-life failures of equipment. One is the probability of an end-of-life failure occurring in a given period (usually one year). The other is unavailability, which is the probability of the equipment being unavailable due to its end-of-life failure during a given period. Both probabilities are used to quantify the likelihood of equipment's end-of-life failure, although they are conceptually somewhat different. The unavailability due to end-of-life failure is consistent with the concept of the unavailability due to repairable failure of equipment. The end-of-life and repairable failures are two basic failure modes in system risk assessment (IEEE, 2006).

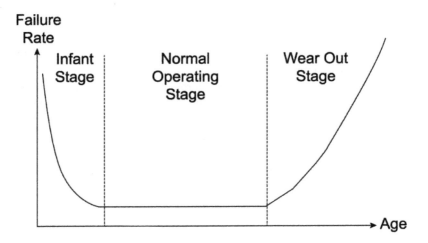

Figure 5.4-1. Basin Curve for Failure Rate of Equipment

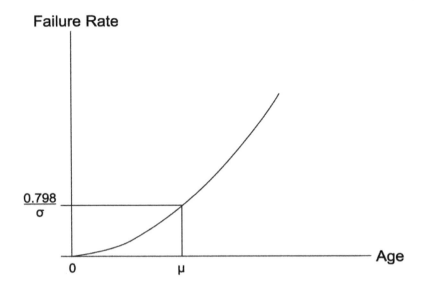

Figure 5.4-2. Relationship Between Failure Rate and Age for a Normal
Probability Distribution

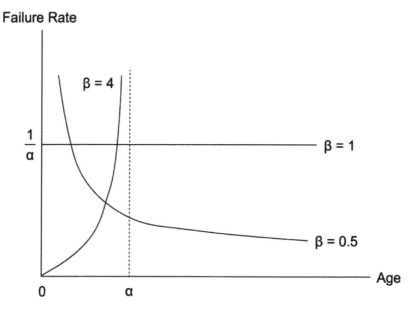

Figure 5.4-3. Relationship Between Failure Rate and Age for a Weibull Probability Distribution

5.4.3   Aging Process and Maintenance

As shown, the probability of the end-of-life failure increases with equipment aging. On the other hand, the progression of the service aging process can be evaluated leveraging the asset integrity program. Conceptually, there are two types of maintenance: reactive and preventative. Reactive maintenance consists of performing maintenance (or repair) after the equipment has failed in service. The actions taken are confined to the specific failure event to restore the equipment to an acceptable level of operation. This type of maintenance is reserved for those assets that do not have a direct impact on the reliability and availability of the overall process. Preventive maintenance consists of performing maintenance activities at predetermined intervals (typically time based) in an attempt to prevent a breakdown. This type of maintenance seeks to reduce the frequency and severity of unplanned shutdowns by establishing a fixed, time based schedule of routine inspections. The major goal of preventative maintenance is to reduce deterioration and prolong the lifetime of equipment, and it addresses both repairable and end-of-life failures. The usable or economic value of equipment is reduced as it ages, and preventative maintenance activities delay the aging process. The relationship between the economic value, time, and preventative maintenance is shown in Figure 5.4-4. It can be seen from the figure that maintenance can recover part of the lost value caused by deterioration in the aging process. However, although maintenance can slow aging, it cannot fully stop it (IEEE, 2006).

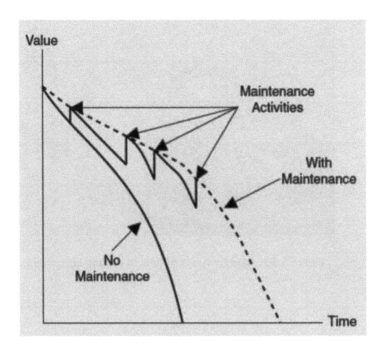

Figure 5.4-4. Relationship Between the Value, Time, and Preventive Maintenance for Aged Equipment

5.5   INFRASTRUCTURE SPECIFIC TOPICS

Chapter 7 contains guidance on managing specific categories of infrastructure. They generally cover the management system aspects of aging assets and the inspections that are necessary to support the management efforts.

Similar to the approach for process equipment integrity management, the facility should develop an infrastructure integrity management guideline procedure. The procedure should state the program purpose and goals, identify what assets are to be covered, and provide general guidance regarding asset design and construction information, and inspection considerations including record keeping. Asset information would include as a minimum applicable codes and standards, design capacity limits, civil, mechanical and electrical drawings as appropriate. The infrastructure specific chapters provide more details on the development of the management systems and inspection support programs.

Figure 5.5-1. Aged Conveyor System in Backup Service

6

INSPECTION AND MAINTENANCE PRACTICES FOR MANAGING LIFE CYCLE

This chapter describes Inspection, and Maintenance (I&M) practices for dealing with Aging Infrastructure. I&M is critical to asset integrity, which is an essential management system for addressing the risks of aging infrastructure assets. It builds on the principles of process equipment asset integrity as they apply to infrastructure assets. Inspection and maintenance practices for dealing with aging assets will only be effective if a formal documented management system is in place and is followed. Aspects of such a management system are presented herein.

Figure 6.1-1. Vintage Steel Mill Retired from Active Service

6.1   INSPECTION AND MAINTENANCE GOALS

6.1.1 Vision

The vision for the kind and, more importantly, the quality of the inspection program for infrastructure, is the providence of senior management. Without management support and commitment, plant level managers will be hard pressed to succeed and will be beset by frustration. Chapter 3 discusses the importance of upper management buy-in and what shape it should take.

Establish a culture of normal excellence – Culture needs to support "best in class" in all aspects of business. Employees who recognize gaps in the vision need to be empowered to act. As a starter, tell employees that you care. If management cares, workers will care. However, management telling workers they care about infrastructure doesn't drive culture as much as actions. If projects related to fixing infrastructure are constantly declined or sidelined, actions are not supportive of a caring culture.

6.1.2 Inspection and Maintenance Commitment for Expected Lifecycle of Equipment

To succeed, the facility needs to commit to maintaining the asset for the designated life expectancy of the facility, which in many cases is 40 years or more. This includes a periodic inspection and monitoring of assets to know when they are reaching a critical point in the lifecycle and require additional attention to prevent significant deterioration and unexpected failure. The commitment should avoid a too common pitfall of deferring maintenance of an asset just because the end is approaching. In some cases, this may be feasible where a plan and schedule is in place for major overhaul or replacement, as long as the condition of the item is closely monitored.

6.1.3 Implementation of Formal Comprehensive Inspection, Testing and Preventive Maintenance Program

A quality inspection and maintenance program requires having formal and comprehensive practices and procedures. This starts by setting goals regarding the intended output of the program, and then designing a program to include practices, procedures, tools, and human resources to meet the goals. Desired program outputs include minor maintenance, repairs, and scheduled replacement of assets when called for. The intent being that periodic minor maintenance will lengthen the cycle before major repairs or replacement is necessary.

Since maintenance and inspection of equipment has been a central activity in the process industries for many years, there is a great deal of guidance available that is applicable to management of aging infrastructure. One body of information that contains RAGAGEP are codes and standards published by professional organizations. Such standards are written by committees made up of member company experts and are peer reviewed. As such, they try to incorporate the best practices from across an industry or profession. There are several Recommended Practices (RPs) that specifically address maintenance and inspection activities such as American Petroleum Institute (API), National

Association of Corrosion Engineers (NACE) International, The Worldwide Corrosion Authority for corrosion, and the American Society of Mechanical

Figure 6.1-2. Vintage Chemical Plant Dust Reduction Facility

Engineers (ASME) for various categories of equipment. A listing of practices that can be applied to aging infrastructure is provided in Chapter 7.

6.1.4   Need Justifiable Inspection and Maintenance Practices

The I&M practices need to be justifiable for a couple of reasons. An important one is to limit liability in the event of an accident involving injuries that was caused by the failure or collapse of aging infrastructure. Having a robust and *documented* program for inspecting and maintaining assets may be needed to counter claims of willful neglect in lawsuits.

A second reason is related to mergers and acquisitions. Usually prior to final approval of a transaction, the facilities are subject to a due diligence review and inspection, or it is a condition in the sale agreement that allows claw back of funds if the equipment is found to be significantly degraded. Having documented inspection and maintenance histories may be instrumental in showing that the asset condition was properly managed and that any deficiency found was not readily foreseeable.

6.1.5   Managing Aging Asset Strategies

There are two general management strategies for addressing the aging of assets. Method 1 determines the remaining life expectancy based on predictive monitoring. This method is strong on monitoring, which dictates follow-up activities and it requires a documented and justifiable program to address asset integrity. Typical components can include but are not limited to:

- Original design specifications
- Service records – asset incident reports, repair history, maintenance performed on asset and co-joined equipment
- Detailed inspection plan
- Visual inspections as minimum
- Rigor of methodology based on RAGAGEP or Failure Modes, Effects and Consequence Analysis (FMECA) for each system
- Non-destructive inspections for metallic piping corrosion, such as ultrasonic testing or magnetic flux leakage surveys
- Radiographic, thermography and other NDT where need is identified
- Need to inspect hard to access regions like insulated lines and lines extending over roadway not easily reached by an aerial lift. Accessibility should not have any bearing. The inspection strategy should be based on the process hazards, damage mechanism and failure modes
- Monitoring land movement that interacts with piping or other structures

Method 2 specifies a target life expectancy, then develops plans and does what is necessary to achieve it. This approach is sometimes utilized by utility companies, because not all deterioration can be identified through inspections. When in-service failure poses an unacceptable risk in terms of cost, reliability, and/or safety, and customers cannot wait for equipment to fail, a utility should replace it preemptively (SCE, 2015). For example, underground cable is unique in that it cannot be visually inspected. Without a deliberate preemptive replacement program, cable would be removed from the system only as a result of in-service failure (SCE, 2015).

To be cost effective, preemptive replacement requires acquiring data on that the average time to wear-out, otherwise replacement may occur too soon. The curve shown in Figure 6.1-3 depicts the relationship between the probability of failure and cable age at one company. This data indicates the failure frequency is low for an average wear-out time of about 30 years.

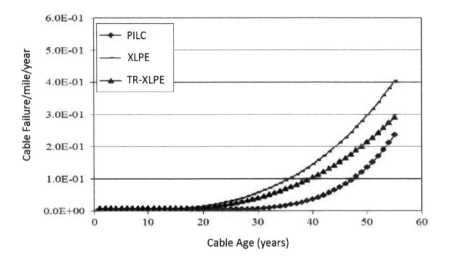

Figure 6.1-3. Cable Failure Rates

Applying pre-emptive replacement is usually not done without some consideration of risk. This is done by evaluating outage records to identify the worst performing circuits in terms of metrics like System Average Interruption Duration Index (SAIDI), System Average Interruption Frequency Index (SAIFI), Circuit SAIDI, and Circuit SAIFI. The most risk significant equipment/infrastructure in the worst-performing circuits is identified for replacement. This requires maintaining records of distribution assets, tracking distribution system and circuit reliability, identifying actual and probable performance trends, in order to draft cost-effective corrective actions where indicated (SCE, 2015).

A similar strategy is used by Targeted Infrastructure Replacement Programs (TIRP) for aging gas utility distribution piping infrastructure, which also incorporates the potential risk of greenhouse gas emissions (PSE&G, 2015).

As described above, Method 2 is more prescriptive and may at first have lower annual operating expenses. However, it may also encounter lack of necessary support as the end approaches and be more susceptible to large unexpected repair costs.

Some examples of infrastructure asset life expectancies are shown in Table 6.1-1. Any user of these values should first refer to the source reference (CIDB) for cautionary notes on their limitations.

6.2 INSPECTION AND MAINTENANCE PROGRAM ELEMENTS

Maintenance and inspection are part of an asset integrity management system. System requirements include:

- Policies, rules codes, standards, practices and procedures

- Inspection and maintenance plan based on the process hazards, damage mechanisms and failure modes.
- Deficiency management process, including repair, alteration, re-rating and replacement tracking
- Asset maintenance records, inspection and service archives for trending of results
- Remaining life / fitness for service assessment
- Training and qualification of maintenance, inspection and testing personnel
- Metrics on success and quality of implementation
- Audits
- Reporting /communication of results to management

These elements will be discussed throughout this chapter and in more detail in Chapter 7.

Type of Infrastructure	Average Annual Maintenance Budget as % of Replacement Cost	Key Assumptions	Replacement or Major Rehabilitation over and above the Annual Maintenance Budget requiring specific capital budget
Bulk water storage	4-8%	Mostly for periodic repair of electrical and mechanical works, storm damage repair, routine maintenance and periodic maintenance	every 30 to 50 years
Water treatment works	4-8%	Mostly for electrical and mechanical equipment	every 20 to 30 years
Water reservoirs	2-3%	Generally low maintenance mostly of telemetry and electrical equipment, storm damage repair, pipe work repair, safety and security, routine maintenance and periodic maintenance	every 20 to 30 years
Water reticulation	4-8%	Mostly for telemetry and pumping equipment, emergency leak repair and ongoing leak repair due to degradation, storm damage repair	every 20 to 30 years
Sewage treatment works	4-8%	Mostly for electrical and mechanical equipment, storm damage and periodic maintenance	every 20 to 30 years
Sewer reticulation	4-8%	Mostly for pumping equipment, emergency leak repair and ongoing leak repair due to degradation, blockage removal, storm damage repair	every 20 to 30 years
Roads and storm water	5-10%	Mostly for emergency repair, storm damage repair, and periodic maintenance (resurfacing every 7 to 10 years)	every 20 to 30 years
Public buildings	4-6%	Mostly for emergency repair, storm damage repair, and periodic maintenance (e.g. repainting and cosmetic upgrades every 5 to 10 years)	every 30 to 50 years
Hospitals	5-8%	Mostly for emergency repair, storm damage repair, and periodic maintenance (e.g. repainting every 3 to 5 years, and cosmetic and operational upgrades every 7 to 10 years)	every 20 to 30 years
Schools	4-6%	Mostly for emergency repair, storm damage repair, and periodic maintenance (e.g. repainting every 5 to 7 years,)	every 30 to 50 years
Electricity generation	5-8%	Mostly for electrical and mechanical equipment and dependent on age and technology of works	every 30 to 50 years
Electricity reticulation	10-15%	Mostly for emergency repair, storm damage repair, safety and security, routine maintenance and periodic maintenance (e.g. every 7 to 10 years)	every 20 to 30 years

6.2.1 Maintenance Program

Coverage. An initial process involves determining what infrastructure assets are going to be covered by the infrastructure maintenance program. Items that are typically included are utilities and support systems, roads, pipe racks and bridges, above and underground piping and cables, marine facilities, free standing stacks, cooling towers, waste ponds, electrical distribution systems and area lighting, fireproofing, drainage, supporting structure, buildings and foundations, and Heating, Ventilation and Air Conditioning (HVAC) systems.

A complete inventory of the targets assets should be documented. The assets should then be assigned a priority rating based on the risk to the continuity of operations which will determine the basis for the type of and scope of maintenance to be performed and the frequency. The criteria for the prioritization should include risk factors addressing process hazards including damage mechanisms and failure modes, safety and operational criticality, environmental impact, and replacement/business cost as a minimum.

Challenges. There is sometimes a tendency for management to view infrastructure costs as a nuisance, because there is no apparent direct line to profits. The challenge is to move the focus from a cost category to a "need to maintain" category. Otherwise maintenance repairs will only be made in response to problems. However, an infrastructure failure can contribute to a major loss. For example, a truck of hazardous material sliding off a road or a delicate shipment being damaged can seriously impact a business.

One approach involves comparing the cost of a potential accident due to lack of proper maintenance (e.g., damage to the assets, public disruption and consequently the degradation of company image and worse fatality) with the cost (net present value) of an ongoing maintenance program for infrastructure. Adding a criterion for impact to company image and reputation to the asset priority ranking process can help incorporate less quantifiable risks.

Another option is to lobby for incorporation of an "other" category as part of plant maintenance program 10-year forecast, annual budget, cost stewardships etc. The intent is to budget for the long term by having funds available for infrastructure repairs that are likely to be needed as assets age.

Maintaining the infrastructure also instills a sense of pride and ownership in the workers, which has subtle benefits that are not quantifiable. A happy worker is a productive worker.

Maintenance has to be a front-end program. Even the best maintenance practices can't erase a bad history of abuse and neglect. Age and neglect can be a big challenge for acquired facility management. When a facility has some doubt about its financial future, there may be a tendency to avoid costly maintenance in the hope that a future owner may take over the operation and assume responsibility for the equipment. On the receiving end, even though equipment maintenance records and archives may exist one has to question the depth and rigor of such maintenance when the future was uncertain. Issues that may not be detected at the time of the sale include hard to detect or repair deterioration like corrosion under fireproofing or insulation, weakened support structures (steel and concrete), and underground infrastructure (e.g., leaking sewers, chemical or waste water drains with environmental consequences).

Policies, Practices, and Procedures. Policies, Practices and Procedures are the tenets that taken together form the foundation of a durable management system. A policy is a broad affirmation of the company's goal on how it will conduct itself. With regard to the I&M program it might include a statement such as "As a company we will not knowingly operate equipment or infrastructure that poses a risk of injury to our employees, the general public, environment or our enterprise." That is a reminder to all management, that once such a risk is known, some mitigating action needs to be implemented.

Practices incorporate the "What we intend to do" to achieve the policy goals. They typically take the form of internal company standards that facilities need to follow in implementing a management program, such as Inspection and Maintenance of Process Equipment and Infrastructure. The standard will describe the scope (what will be covered by the program), criteria for setting I&M priority, the types of inspections and preventive maintenance, and what association RAGAGEPs or company practices to employ, information management systems for work orders, maintenance and inspection records, and other documentation, skills and training requirements and management reporting.

As previously explained, RAGAGEP includes consensus practices that have been found to be effective for some kinds of integrity issues generally found in the process and other industries. RAGAGEPs do exist for facilities and infrastructure; however, company/site policies may need to be developed. As such they are good references for developing a basis for a sound I&M program. Since the issuers of these codes and best practices cannot anticipate all the conditions that may exist in a user's facility or its infrastructure, they should not necessarily be the only methods employed to monitor integrity. Plant and company operating experience on similar systems should also be considered.

Finally, there are written procedures that address how people tasked with the responsibility of implementing the program requirements are supposed to do their jobs. For example, take the visual inspection of a pipe rack. It would begin with how the work order is generated and assigned to an examiner. Next the examiner needs to know how to obtain a copy of the inspection reporting form and instruction (e.g., from a computer based system). Examiners then need to review available inspection history to compare the current status of the rack to when it was last inspected. The field inspection may need the maintenance department to provide access to the asset which needs to be scheduled. After completion of the field examination, the examiner inputs his/her report into the documentation system.

The system may then communicate with the inspector assigned to the asset to notify that the field inspection is complete. The inspector will access the report and review to see if there are any issues that need to be addressed. He/she may contact the examiner to obtain additional information. If the condition of the asset is such that an action is required, the inspector decides what needs to be done, whether it qualifies as a normal maintenance expense, and initiates a work order to initiate the work. If not, he/she will need to get authority from the maintenance manager on how to plan for and schedule the repair. The procedure should explain these steps and include examples of forms that are required.

A well implemented I&M program will have a hierarchy of written documents of the type presented above, that provide the substance of the program and what is expected.

Justifiable and Documented Program. Developing a justifiable documented program means paying attention to details and being rigorous. It starts with understanding the current operation and determining how equipment and facilities are matched to it. Verify whether safe operating limits and/or IOWs have changed since equipment was first installed and why. Investigate if there is any evidence of aging changes occurring under present conditions.

To achieve a justifiable program, documented information is required about the design, construction/fabrication, operation, maintenance and repair history, and current condition of assets. The design information category should contain the original equipment or asset records, including both process design specifications (if appropriate), and civil, mechanical and/or electrical design data depending on the asset type. The purpose is to define the starting point in the asset lifecycle. For older facilities, this is more problematic due to loss of records, especially if the ownership of the facilities has changed hands. In that case, the baseline must be established using the earliest date when inspection measurements were made.

It is important that the operating history is recorded whether changes in the service or operating conditions have occurred since the asset was commissioned. Facilities which operate in a process safety regulated jurisdiction may have management of change records, but these may not include all infrastructure assets. For infrastructure assets, maintenance/repair records may pin point when something was added or upgraded and the reason. This information is used to compare current or future service, load or capacity conditions to the original design conditions and safe operating limits. At some point in an asset's lifecycle, such a comparison may indicate the need for a fitness for service evaluation.

A key component of the maintenance program documentation is the preventive maintenance and trending results and repair/alteration/re-rating/replacement records. Each covered asset should have a database repository that documents the preventive maintenance regime required, task instructions, the dates performed, and any anomalies found and adjusted. The task instructions may include the Preventive Maintenance (PM) scope, type & tools needed, schedule, asset location and accessibility information, and any forms that need to be used to record the PM completion. When anomalies are found that are not correctable by routine maintenance, the system should notify the inspection department to investigate the condition of the item.

The system should also document the date of all overhauls or repairs done to the asset, a brief statement of the reason for the action (e.g., sanded & repainted failed coating or, removed and replaced cracked and spalled fireproofing) and the Work Order (WO) number. Details about the repair may be recorded in the work order system.

Keep in mind that at some future date someone else may be asked to review these records, and it could be the company's legal department and the plaintiff's expert witness. This can happen when an incident with casualties occurs or after an acquisition of assets that were warranted as fit for service by the seller, but later found defective.

Condition Monitoring of Infrastructure. W.E. Demings' classic statement "You can't control what you don't measure!" can be rephrased for asset integrity "You can't manage what you don't monitor!" This brings to mind another well-worn adage "Ignorance is no excuse when it comes to the law!"

Condition and performance monitoring are essential to help verify integrity to proactively manage infrastructure aging and potential serious or catastrophic incidents. Fortunately, today there are methods and tools available to engineers that can be applied to assist with condition and performance monitoring. The next section of this chapter includes discussion on techniques that are available for various types of infrastructure and failure mechanisms including many non-destructive testing methods.

Performance Metrics. Managers with responsibility for implementing and guiding the maintenance program need to answer the question "Are we doing what we say we are doing, and how do I know?" Developing suitable metrics is a way to address that question. The metrics need to focus on two aspects:

1. Is required maintenance being performed and on schedule?

2. What is the quality of the implementation?

Some typical Key Performance Indicators (KPIs) for monitoring maintenance programs are summarized below with a brief explanation of their purpose.

Preventive maintenance compliance (leading). This indicator is expressed as percentage of PM work orders completed compared to the total PM work orders due on a monthly or quarterly basis. High values for on-time completions indicates effective maintenance planning and execution. (Competed PM WOs/Total due PM WOs *100)

PM corrective work (leading). The ratio of Corrective Maintenance (CM) work needed as a result of PM activities compared to the amount of PM work being performed is a measure of the effectiveness of the PM program to reduce repairs. The ratio value should be low and preferably declining. (CM work/ All PM work)

PM versus All Maintenance (leading). The ratio of PM work activity to all maintenance work activity should increase as activity shifts away from breakdown maintenance towards preventive maintenance. A useful variant of this concept is the ratio of PM activities to CM activities for a specific asset. A declining ratio indicates the asset is far into its lifecycle and the optimum (cost effective) time for replacement may be approaching. (PM work/All Maintenance work)

PM Work Backlog Trending (leading). The objective of this indicator is to manage the PM work order backlog. This indicator is a measure of all active PM work orders in the system. It is historically trended using the required due by date of the work order and comparing this to the current date +/- 14 days. Using this guideline, all active PM work orders are segregated into categories 'Overdue', 'Current' and 'Future', according to a predetermined calendar based formula, and plotted as a function of time. The graphical representation allows the maintenance manager to identify trends in non-compliance and effectiveness of backlog reviews (EUR 22602 EN, 2006).

Overdue PM Work. The proposed indicator is a measure of PM work orders that are past the required due date (i.e., overdue). It can be expressed as a percentage of overdue PM work orders to the total PM work orders due each month. Alternatively, each overdue PM work order can be assigned a percent overdue =

[(Current date – Required due date) / PM frequency (days)] x 100. The PM work orders that are determined to be overdue can then be rank ordered to determine those that are the most overdue, and need immediate corrective action. Another variant is to track schedule PM work order compliance as a percentage of all PM work orders due in particular month, with comparison to a benchmark target (>90%). High values of overdue PM work are evidence of poor planning or inadequate attitude of plant management.

Emergency Maintenance (lagging). This is one of the more concrete, albeit after the fact, measures of the health of the maintenance program. No one, especially management likes surprises. When the percentage of Emergency Repair (ER) work orders starts to increase above a target benchmark, it is signaling the maintenance program is not adequately performing its mission. It should trigger a review to identify where and how preventive maintenance and inspection practices and implementation need to be improved. (ER WOs/All Maintenance WOs*100)

In 2010, the US Department of Homeland Security organized a workshop on Aging Infrastructure: Issues, Research, and Technology (US DHS, 2010) which resulted in some suggestions for metrics for aging infrastructure. A few of the findings are presented below:

- A combination of criticality and vulnerability of assets can be used as a prioritization metric
- The probability of events might be a useful metric in certain conditions
- Single points of failure should be accounted for in establishing metric
- Risk is a popular metric that should be used and is generally defined. When used as a prioritization metric, consequence should be considered as loss of asset or denial of service, and should be based on public health/safety, socioeconomic impact, and environmental impact
- No reliability analysis is available for infrastructure needs, so a method would need to be determined
- Use current national code and standards to develop pertinent metrics. Those codes were developed based on rigorous engineering criteria, they can be valuable and accurate prioritization metrics

These offerings also provide insight into the problem of prioritizing and addressing the nation's aging infrastructure.

Auditable Program. Proactive companies typically perform periodic audits of key management systems (e.g., Asset Integrity) to verify compliance with applicable regulations and company standards and practices. The principal purpose of these audits is to identify gaps and continually improve the program.

The section on developing a justifiable documented program explains that when there is good documentation the auditing process is simplified and can drill down to quality aspects of implementation, rather than only trying to determine if all program elements are covered. The plant benefits in that the finds are more specific and actionable, rather than simply restating known gaps. Generally, the audit will require less time on the part of the plant personnel when the auditors can quickly access documents and records for verification.

When there is a significant incident at a plant that impacts employees, regulatory authorities will often investigate. Once at the plant, they are not limited to just investigating the causes of the incident. Any management system that may affect workplace safety (such as asset integrity), is within the scope, and the auditors might want to see documented evidence to verify compliance with regulations, as well as associated company standards and RAGAPEPs.

Management Reporting. Facility physical asset integrity is a serious matter that plant and regional managers must be involved in. By developing a practice of requiring metrics as described in section 6.2.1.6, preparation of key performance indictors becomes a routine activity each month. This information should be compiled in an executive summary for management review on a periodic basis (e.g., monthly for plant level, quarterly for corporate level). This will allow senior management to track performance and progress on meeting benchmark targets, and propose adjustments if performance trends are deteriorating.

6.2.2 Inspection Program

Inspection is the other major pillar of asset integrity management, alongside maintenance. To a large extent, preventive maintenance is a lagging action dealing with the current conditions of assets, or at shorter term horizons (to next due date). Inspection programs can be leading or forward looking in scope with longer term views. Inspection coupled with trending is an essential feature of predictive maintenance systems.

As the saying goes, "EXPECT WHAT YOU INSPECT". That is to say design an inspection program to find the types of aging and degradation that you believe can occur, and then be prepared to find it. Absence of visible defects may not provide sufficient proof that a future failure is not imminent. Make sure inspections are thorough and not skin deep. Look for the unusual! Do not only inspect physical assets, but also inspect human assets, i.e., the knowledge of the operators, information in operating manuals and how asset changes are communicated between operators.

The initiation of the inspection and verification processes should be implemented as soon as the asset is installed and continue throughout its lifecycle. This includes establishing initial base line conditions and attention to accurate record keeping to allow proper trending of changes in condition.

Most of the topics discussed in section 6.2.1 Maintenance Program apply equally to inspection. Inspection program metrics will vary somewhat and are discussed later in this section.

Infrastructure Aging Indications. Although some aging indications are not easily seen, visual inspection of assets is a viable method at least for a preliminary inspection of above ground systems. The objective of a preliminary inspection is to obtain initial analytical information to assess the physical adequacy of an existing asset. This inspection is usually a field examination of the asset to visually evaluate the structural components.

The visual inspections should be thorough and not rushed. Be prepared to make some physical measurements (e.g., crack width and depth) and take photographs of warning signs. In this case, ugliness may be more that skin deep, look for indications that point to an underlying problem (e.g., rusting stains on the outside of reinforced concrete or fireproofing). There may be some surface indications of deteriorated pressurized underground cooling water and firewater piping such as wet or sinking soil.

Table 6.2-1 and the tables in Chapter 7 provide some examples of signs of specific infrastructure aging to look for when performing a visual inspection.

Table 6.2-1. Example Checklist for Maintenance and Inspection

MECHANICAL - PUMPS AND COMPRESSORS	Y	N	NA	Remarks
1. Is there excessive packing leakage or leaking seals?	☐	☐	☐	
2. Are there loose support bolts or cracked foundation pedestal?	☐	☐	☐	
3. Are there loose or missing equipment guards?	☐	☐	☐	
4. Are there squeaking noises or excessive vibration?	☐	☐	☐	
5. Is there protective coating degradation (such as discoloration, blistering, cracking, peeling, or dissolving) or insulation damage?	☐	☐	☐	
MECHANICAL - VALVES	**Y**	**N**	**NA**	**Remarks**
6. Is the exposed stem of any valve corroded?	☐	☐	☐	
7. Is there missing or loose hand wheels, chain wheels, or lever arms?	☐	☐	☐	
8. Is there corroded valve trim?	☐	☐	☐	
9. Are there bent, broken, or missing valve position indicators or limit switches?	☐	☐	☐	
10. Are there safety wire seals broken on relief valves?	☐	☐	☐	
11. Is there leaking flange joints or packing?	☐	☐	☐	
MECHANICAL - PIPING CONDUIT AND ANCHOR	**Y**	**N**	**NA**	**Remarks**
12. Are there flange nuts, studs, or bolts that are missing or not fully engaged?	☐	☐	☐	
13. Are there corrosion stains seeping through thermal insulation or wetted thermal insulation?	☐	☐	☐	
14. Is there temporary shielding suspended from pipe?	☐	☐	☐	
15. Are there cracked or deformed elastomeric expansion joints?	☐	☐	☐	
16. Are there plastic tie wraps or wire supporting pipe?	☐	☐	☐	
MECHANICAL - PIPING AND ANCHOR	**Y**	**N**	**NA**	**Remarks**
17. Are sight glasses visible?	☐	☐	☐	
18. Are there any visible vibrations?	☐	☐	☐	
19. Paint failure and contact point failure?	☐	☐	☐	
20. Insulation missing?	☐	☐	☐	
21. Water ingress points?	☐	☐	☐	
22. Sagging of piping?	☐	☐	☐	
23. Are there any tubular dummy legs (unseen corrosion)?	☐	☐	☐	
24. Are there any shoes off their supports?	☐	☐	☐	
25. Are the spring hangers bottomed or topped out?	☐	☐	☐	
26. Do the spring hangers have any gages that should be out?	☐	☐	☐	
27. Are there flange nuts, studs, or bolts that are missing or not fully engaged?	☐	☐	☐	

Table 6.2-1. Example Checklist for Maintenance and Inspection, continued

MECHANICAL - PIPING AND ANCHOR	Y	N	NA	Remarks
28. Are there corrosion stains seeping through thermal insulation or wetted thermal insulation?	☐	☐	☐	
29. Are there temporary shielding suspended from pipe?	☐	☐	☐	
30. Are there cracked or deformed elastomeric expansion joints?	☐	☐	☐	
31. Are there plastic tie wraps supporting pipe?	☐	☐	☐	
MECHANICAL - TANKS	**Y**	**N**	**NA**	**Remarks**
32. External floating roof-roof drain functional? Look up tank inspection procedure.	☐	☐	☐	
33. Tank vents inspected and unobstructed?	☐	☐	☐	
34. Any damage on tank pad or ring wall?	☐	☐	☐	
35. Is cathodic protection in place and working?	☐	☐	☐	
36. Is there any shell distortions or signs of settlement?	☐	☐	☐	
37. Is there corrosion product on exterior of tank wall, or man way bolting?	☐	☐	☐	
38. Is there any damage on tank pad?	☐	☐	☐	
39. Are there any cracks on dike wall?	☐	☐	☐	
40. Are there wet signs on tank bottom proximity?	☐	☐	☐	
41. Is there protective coating degradation (such as discoloration, blistering, cracking, peeling, or dissolving)?	☐	☐	☐	

Inspection Program Metrics. Key Performance Indicators (KPIs) are utilized to gauge the effectiveness of important actives and endeavors and are predetermined, quantifiable measurements that reflect the critical success factors (for example a program). The acronym Specific Measurable Attainable Realistic Timely (SMART) is used to characterize the requirements of a good KPI. Some typical KPIs for inspection programs include:

- Number of overdue inspection work orders
- Number of high priority inspections overdue more than 30 days
- Number of inspection repairs vs scheduled repairs
- Percentage of planned inspection program completed

The following KPI is a measure of the overall cost effectiveness of the inspection program.

- Preventive inspection effectiveness = (Preventive repair man hours / Preventive inspection man hours) x 100%
- Preventive repair man hours refers to maintenance that is performed as a result of a preventive inspection

- Preventive inspection man hours refers to work performed that originates from equipment maintenance strategies, i.e., planned inspection services and inspections

These KPIs can be combined with the ones defined in previous sections as a measure of the overall maintenance and inspection program.

6.3   INSPECTION AND MAINTENANCE PROGRAM RESOURCES

There is a need to have the right resources to support the maintenance inspection program.

- Skilled and certified examiners and inspectors
- Skilled mechanical/metallurgical engineers
- Competent contractor base (NDE examiners) using reliable inspection equipment
- Senior leader or manager to champion the program and lobby for additional funding when necessary

6.3.1   Human Resources

Competency. Employees are a company's most valued asset. The first line of defense for managing infrastructure asset integrity is the knowledge and competency of people engaged in the program. It starts with assigning personnel to key positions that have the required skills and experience. If assignments in the maintenance and inspection department are leadership positions for promotion to more senior management, then the candidates should be properly trained, if they do not already have all the necessary technical background needed. There are many examples of managers being assigned to a department for a year or so to get experience and then moving on. This may be good for the employee, but not for the continuity of the department programs. Maintenance and inspection are serious technical activities, which need knowledgeable and dedicated people to implement.

For staff that desire to remain in I&M for the duration of their careers, it is necessary to maintain their competency. Make sure they are conversant with new technology and trends and are not just doing what they always have done, if there are better methods. Maintainability of technical skill base is important. Asset integrity should be part of everyone's responsibility involved in operations, maintenance, and technical support. Target a percentage of annual hours for refresher training. A possible minimum allocation is 5%, but each company should set a target that is appropriate.

Finally, but not least, employees who recognize gaps in the program need to be empowered to act, write work orders, submit loss reports, and challenge the status quo. The message should be "Don't let the place fall apart". Also, don't leave out the operators. If operators are engaged in condition monitoring (they see the assets every shift) and are responsible for issuing work orders, more attention can be brought to bear on deteriorating infrastructure issues.

Certified Examiners and Inspectors. For companies applying *API Standard 510 Inspection Practices for Pressure Vessels, RP 570 Inspection Practices of*

Piping, and RP 653 Inspection Practices for Above Ground Storage Tank, examiners and inspectors need to have appropriate education and years of experience, pass an exam, taken a certification training course and receive recertification every 3 years. Every 6 years, certified inspectors must retest to validate they are aware of recent inspection codes. Per API, examiners are the field personnel that support inspectors actually in performing the inspection of the equipment. Inspectors are qualified to interpret the results of the equipment condition reports and recommend further actions including additional testing, PM repairs or fit-for-service analysis.

While many infrastructure assets are not specifically covered by the API practices listed above, the need for qualified personnel performing the examinations and interpreting the results is still valid. They should be knowledgeable of the visual and other inspection methods appropriate to the type of infrastructure being inspected using practices and guidelines in references like the ones described earlier. Even though equipment specific RAGAGEPs may not exist, there is RAGAGEP on qualification and certification of personnel conducting visual inspections and other inspection methods through American Society of Non-destructive Testing (ASNT) and other organizations.

This does not suggest that the examiners and inspectors need to be different than the ones with API certificates. In fact, there are many benefits for using the same people. They already know their jobs and some of the same principles and inspection methods apply to certain assets (e.g., steel structure, foundations). They may need to brush up on some NDT methods for certain types of assets like determining the condition of re-bar in reinforced and precast concrete. A note of precaution, in a RAGAGEP driven environment (such as process safety), certified API inspectors may be reluctant to inspect equipment outside the scope of their certifications without training or certification on other applicable RAGAGEP. The main consideration is that infrastructure assets are included in a comprehensive I&M program and are assigned to qualified inspectors.

Maintenance and Inspection Engineers. While infrastructure assets may or may not be directly involved in handling process materials, their integrity is a technical enterprise that requires involvement of engineers. The usual engineering disciplines include mechanical, electrical, civil and metallurgical. Generally, an individual has an accredited degree in one or more of those disciplines and sufficient acquired knowledge in another through professional association courses.

Experience in maintaining and inspecting equipment is also an important aspect. One pitfall to be aware of is how recently graduated engineers are utilized. It is common and appropriate practice to assign certain new engineering graduates to the maintenance or inspection department. There are a couple of issues with what the individuals are tasked to do and how they are supervised, and how long they are scheduled to stay in that department. The following case study illustrates some of these problems.

Due to a high backlog of unrecorded inspection reports, newly graduated mechanical engineers were given the responsibility of entering and reviewing the reports and determining whether further inspection or maintenance was required. Expanding the scope of an inspection is normally the responsibility of an API certified inspector. The young engineers were not certified and not properly supervised by certified personnel to undertake this responsibility.

The recent graduate engineers were in a company Engineer-in-Training (EIT) program, and only stayed in the department for a few months.

Consequently, they were not there long enough to gain sufficient experience or certification to be proficient with the job. In the end, the quality of the I&M program was compromised due to lack of appropriate recommendations for additional field examination or for immediate easy repairs that would have retarded the aging process. Because the I&M program had not received the necessary attention afforded by an experienced inspector at the right time, some equipment eventually required extensive repairs following a FFS assessment.

Contractors. Invariably, I&M programs utilize contractors for maintenance work and specialty services such as PM of pump seals, equipment cleaning, and inspections employing nondestructive testing methods. However, the I&M management is still responsible and the effectiveness of the program is only as good as the weakest link. Therefore, it may be necessary to oversee and periodically audit the quality of contractor's performance.

For contractors that require certification, this involves making sure their certificates are current and up to date. Occasional observation of contractor work by a company supervisor should be performed. For NDE contractors, understanding how and when the equipment is calibrated should be determined and audited. Contractors should be required to meet the same exacting standards as any company employee involved with I&M, and they should be informed that is what is expected.

Senior Leader/Manager. Much has already been said in this book and others about the importance of management setting the example. General Patton (a famous US army commander in World War II) was, by all accounts, a hard task master. But his troops respected him and literally died for him. One characteristic he exuded in large measure was passion.

When auditing company asset integrity programs, one thing that stands out is that the best-in-class programs usually have a champion who demonstrates passion for the work and organization he/she is managing, and can instill that passion to the workers. They are good organizers and pay attention to details by:

- Setting up robust documentation systems
- Requiring progress feedback from direct reports
- Developing and rigorously applying KPIs
- Reporting on performance frequently to management
- Supporting their personnel by providing opportunities to improve their knowledge and skills to maintain best-in-class excellence

Overall, there is very little that goes on in the program that they don't know about. By their example, this sends a message to employees, that the expectation is nothing less than the same commitment. Also, when management is asked for additional resources for some activity or management system improvement project, they expect that there will be a solid reason and good justification, which makes the selling easier.

6.4   ADDRESSING INFRASTRUCTURE DEFICIENCIES

Let's return to the prior statement, "Expect What You Inspect (for)". At some point it may become necessary to address deficiencies that require more than simple maintenance repairs. This section deals with situations when major refurbishment or replacement of an asset is required.

6.4.1 Inspection Follow-up

The purpose of the I&M program is to deal with issues when found, not just merely recording bad results. However, follow-up needs to be systematic and accurate before any decisions and investment in fixes are made. Understanding the causes of integrity issues before fixing is key to avoiding reoccurrence.

Understand Operation and Service Conditions. When it becomes apparent that the asset is going to need significant repairs, it is important to understand how the equipment reached this condition. This involves reviewing the operating and service history as far back as records allow. Especially interesting are any changes that may have been made in the service history, which may have influenced the aging process.

For infrastructure assets, changes in environmental factors can be most significant. The following is an excerpt from the US Department of Homeland Security (DHS) workshop on aging infrastructure.

> "Environmental factors can reinforce or perhaps override age as a contributor to infrastructure failure. Examples of environmental factors often cited as affecting underground infrastructure include soil movement and pressure created by seasonal freeze-thaw cycles and attack by biological or chemical agents in the underground environment. Other environmental factors related more to human actions include construction interference involving inadvertent breakages of utility lines (backhoe failure), failure to back fill supporting material for other infrastructure after construction, and breakages in water lines during winter months that can cause freezing of water around other utilities lines. Infrastructures that are in poorer condition due to age can be more vulnerable to such environmental intrusions. A wide range of other environmental factors affect above ground infrastructure facilities that are weather related and also involve destruction by animals and birds" (US DHS, 2010).

For process facility environments, factors can also include emissions and other conditions such as cooling tower drift, fugitive and accidental releases, salt air, and interactions from failure of other systems. That means the investigation should look beyond just the operating records of the specific asset, and include location and maintenance records of co-located assets and incident reports.

Compile and Analyze the Data. The I&M data is the next place to focus the investigation. What can be deduced from the data about the aging process? What are the indications that warrant concern? Did the deterioration occur gradually, or was there a seemingly unexplained increase in the near term that needs to be further investigated? Is the change real or an anomaly in the data? Once the data are compiled, verified, accepted, additional systematic analysis techniques can be undertaken.

Root Cause Analysis (RCA). The value of doing an RCA is that it may identify a primary cause which is not directly related to the failure mechanism at all. Take for example a pipe or structural column that was found to have significant corrosion under a covering such as insulation or fireproofing. We understand the mechanism, intrusion of water behind the insulating barrier, continuous moisture held against the metal surface and access to oxygen causing rapid corrosion under certain favorable temperature conditions. So, what do we do? We ensure that water doesn't enter the barrier.

But there may be another basic cause that allowed the condition to reach a point where a major repair was required. By applying root cause analysis, it was also determined that there was no comprehensive guideline for inspecting for corrosion under insulation and furthermore, examiners were lacking awareness and training on CUI warning signs. The field inspection form didn't have an instruction to check for CUI. Also, inspectors were less than fully trained on how to do follow up inspections when warning signs were found. Without addressing the administrative control deficiency, the aging problem could recur.

One of the values of root cause analysis is that it can drill down into underlining administrative controls that may be missed by only focusing on engineered controls. One of the simplest RCA techniques is known as Five Whys. It starts by asking: Why did such an event occur? That usually elicits an answer that is the apparent cause. That cause is then met with that second Why and so on until a true root cause is found. Usually it doesn't take asking "Why" five times to find the real reason for the failure. There are more detailed and systematic methodologies available, which may require an experienced facilitator to conduct. For many situations, Five Whys is sufficient. It is good practice to use RCA when troubleshooting any problem to get the right problem solution.

FFS Evaluation (RAGAGEP, regulation compliance). Just because something is inspected does not mean it is fit for service. The inspection results only show that changes are taking place in the condition of the asset. Depending on the seriousness of the inspection results, assets may eventually need to be evaluated for fitness for continued service. Fitness for Service (FFS) is the ability of a system or component to provide continuous and reliable service while meeting all safety regulations until the end of a specified time period. The FFS evaluation needs to determine whether the asset is capable of supporting the original design intention service through the remaining lifecycle.

The burden of proof rests with the owner/operator to demonstrate that a facility or unit assets are fit for service and safe to operate. In turn, this may require a comprehensive inspection that meets the scrutiny of regulatory agencies. Ensuring that data is available for every static asset to support a fitness for service inspection program is fundamental, because a FFS evaluation is an engineering analysis that requires data on physical parameters over time.

The principle RAGAGEP for FFS is API 579-1/ASME FFS-1, (API 2007) a comprehensive consensus industry recommended practice that can be used to analyze, evaluate, and monitor equipment for continued operation. The main types of equipment covered by this standard are pressure vessels, piping, and tanks.

The material presented in the API Recommended Practice describes how the disciplines of stress analysis including finite element analysis, materials engineering, and nondestructive inspection interact and apply to fitness-for-service assessment. The assessment methods are intended for application to pressure vessels, piping, and tanks that are in service. However, the principles involved are generally applicable to systems under stress and are constructed of metal.

The availability of FFS standards for infrastructure per se is more limited as is evident when performing an internet search. One reason is one size does not fit all due to the diversity of infrastructure types. Some universities are working on monitoring and analytical models for specific infrastructure. For example, Dr. Robert Connor led a group of researchers from Purdue University's S-BRITE Center in completing a Fitness for Service evaluation of the tie girder welds on the Ohio River Sherman Minton Bridge, as well as a robust, remote monitoring program where member stress, pier tilt, wind speed and direction, and ambient and steel temperature data were collected over several months. The full details for the FFS are provided in Appendix Y.

For reinforced concrete structure, the American Cement Institute (ACI) has three publications pertaining to life expectancy:

- ACI Committee 365, "Service-Life Prediction," ACI 365.1R-00, 2000
- ACI 207.3R-94: Practices for Evaluation of Concrete in Existing Massive Structures for Service Conditions, Reapproved 2008
- ACI 364.1R-07 Guide for Evaluation of Concrete Structures before Rehabilitation, 2007

There are also some publications by the Governmental Nuclear Regulatory Commission (NRC) on aging nuclear concrete structures that are referenced at the end of this chapter.

Consider standards and recommended practices for FFS a minimum hurdle to achieve. The company's tolerance for risk may dictate applying more safety factor than determined by the FFS assessment.

Addressing Deficiencies. At some point in an asset's lifecycle, it may become necessary to rehabilitate the asset. The options for addressing deficiencies are limp along with short term fixes, make long term repairs, or replace the asset altogether.

Short Term Fix vs Long Term Repair. Short term fixes may be necessary to address immediate problems and current business needs, but should not be the final remedy for known deficiencies that can affect the fitness for service. Short term fixes generally result in the return of issues later. It is a delay strategy and not a sustainable strategy. The rehabilitation should aim to prevent the causes and mitigate the risk of failure. Actions taken should prevent recurrence. Don't take short cuts. Do it correctly, or do it twice!

As discussed above, always understand an integrity problem before trying to fix it. Avoid fixes that react to symptoms instead of causes. Repairs should address the underlying causes. Don't merely repair a section of localized corrosion in a line when the entire line is showing indications and may require a material change. The philosophy should be to prevent recurrence, do it right once or do it twice!

Some examples of short term fixes that don't properly address permanent integrity issues:

- A pipe leak or thinned wall secured with a clamping tool device
- Repairing torn insulation on a line by caulking to prevent further water ingress without removing the moisture underneath

- Continuing to plug heat exchanger tubes, instead of determining the root cause, replacing the tube bundle, and implementing injection of a corrosion inhibitor

Before opting for short or temporary fixes, it is important to really understand the risk and determine whether the company is prepared to accept it? If temporary fixes fail can you live with possible outcomes and are you prepared to deal with them? Is the potential outcome short-term business interruption or worse, employee injury as a result of a temporary repair failure? In this case, long term due diligence by maintaining reliable equipment and good recordkeeping should be the preferred course.

Naturally there will be situations when "temporary fixes" are unavoidable, e.g., to mitigate emergencies, appropriate maintenance/ repair materials are not readily available. In this case, temporary exemptions for short term repairs can be acceptable. When there is a need to postpone for a pressing permanent repair, there should be limits and rules for that duration and tracking. This situation is usually handed through the management of change procedure.

Non-intrusive Actions. In some situations, an asset can be rendered fit for service by reducing service conditions and rerating the equipment. This practice is allowed for pressure vessels by following certain requirements. Opportunities involving infrastructure may be more limited. One example: for support and spanning structures, placing load limits (like on highway bridges) may be possible. Setting the limits should be based on a technical evaluation utilizing verified condition and physical data and generally recognized engineering analysis methods. University mechanical and civil engineering departments are one resource to consider.

Repair to Achieve FFS. Achieving fitness for service is the goal of repairs to defective aging assets. As stated in Chapters 1 and 2, an asset can be either functionally or physically deficient. Assets that are prone to functional aging (e.g., obsolescence) include some electrical components, and instrumentation and control hardware. The options for achieving FFS for these assets may be limited to partial replacement of components or total replacement of a system (conversion from analog to digital devices).

For physically defective assets, repairs for restoration to FFS should be dictated by the type of asset and cause. For example, for corroded steel support structures, this might include weld overlays, or adding weld plates or gusset plates. Failed fireproofing, which is susceptible to ingress of moisture and corrosion under fireproofing (CUF) can be removed and replaced with a fixed water application system.

For reinforced concrete structures and slabs, this may involve grinding out and repairing cracks, or for more severe internal damage, hammering out the concrete, replacing corroded re-bar, and using concrete that is resistant to environmental attack.

Long-term options for fixing corroded steel or leaking nonmetallic underground piping, may be insertion of suitably rated plastic pipe, or angioplasty cementing of leaking joints in sewer lines.

The overriding consideration being when repairs or replacement to meet the requirements of FFS are required, selecting an option that also addresses the degradation causes (to avoid reoccurrence) is overall cost effective.

Repair vs. Replacement. Repair vs replacement is always a decision to be made when rehabilitation of an asset is required. Figure 6.4-1 shows an example of infrastructure awaiting to be replaced.

Figure 6.4-1. Vintage Grain Elevator Awaiting Renewal (or Refurbishment)

A senior project report at the California Polytechnic State University (Gage, 2013), contains some useful insights regarding this problem starting with the main reasoning for considering replacement.

"There should be three main reasons why equipment is considered for replacement. The first reason is the equipment is depleted of function. A very common example is oil wells. Once there is no more oil in the ground, the well is depleted. In the case of Company X, this would be considered a piece of equipment that is run-to-failure. These items are low cost reliable equipment like small pumps or fans which either have redundancy or can easily be replaced and are not in critical systems.

The next reason for replacing equipment is equipment obsolescence. The best example of this is a computer. Older computers are much slower and have fewer features than their modern counterparts. In addition, older computers are harder to maintain because replacement parts and qualified technicians are much more difficult to find. Obsolete equipment for Company X would include manually operated machining equipment. This equipment could be replaced by Computer Numeric Control (CNC) equipment with better tooling, higher accuracy, consistent precision, and more automation. The safety systems in CNC equipment are also significantly better than those in manually operated machinery.

The last reason for replacement, and also the most frequent, is deterioration due to aging. Any mechanical equipment faces this problem, including cars, airplanes, and bicycles. For Company X, this includes water pipes, cranes, boilers, chillers, ventilation systems, lighting, high bay entrances, chambers, and almost any type of equipment which does not meet the criteria for the previous two reasons.

Even with regular maintenance, the cost of maintenance for these items eventually exceeds the cost of replacement" (Gage, 2013).

The study also describes the Defender-Challenger methodology for replacement decision making. To perform an economic analysis for the replacement decision, there needs to be consideration for the existing piece of equipment and any possible replacements. A common model for this analysis is known as the defender-challenger model. The defender is the existing equipment on the property which is in operating condition. The challenger is the best alternative which can be purchased and installed on site. There is a group of challengers for each defender, these challenges are evaluated independently against one another using incremental rate of return analysis to determine the best challenger.

For all comparisons between the defender and challenger, the Expected Uniform Annual Cost (EUAC) is used in the analysis. The EUAC is calculated by spreading the maintenance and replacement costs across the expected life of the equipment. Equipment that is kept for a shorter time frame has a higher loss in capital value but lower maintenance, repair, and operating costs.

The longer the equipment is kept, the depreciation of the capital value is lower on a per year basis but the maintenance, repair, and operating costs rise. The graph of the total EUAC forms a curve as seen in Figure 6.4-2.

If the defender cost data is available and its EUAC is decreasing, the comparison is between the minimum defender EUAC and the minimum challenger EUAC. If the EUAC is increasing, the comparison is between the defender EUAC for the upcoming year and the minimum challenger EUAC. If the data is not available, an estimate of the information over the remaining useful life of the defender is used instead. (Gage, 2013).

Note that estimating the EUAC for capital requires an assumed discount rate or cost of capital. Also, the minimum point on the Total EUAC curve is the economic useful life of the asset. This is not the same as the total useful line which is longer.

While lifecycle cost analysis is an important factor in deciding when to replace an asset, it is not the only one that should be considered. Some other important considerations used by the Bonneville Power Administration include:

- An asset is near or beyond its expected life
- The asset reliability and the consequences of failure poses an unacceptable risk
- The repair/refurbishment costs exceed the lifecycle cost of an asset replacement
- The asset's performance has been unacceptable and corrective maintenance measures will not lead to acceptable performance

- Additional asset capability is required and the replacement equipment provides that

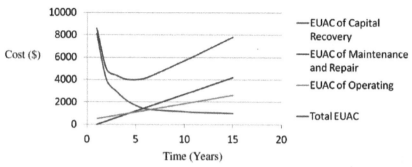

Figure 6.4-2. Expected Uniform Annual Cost

- The existing equipment is technologically obsolete, spare parts are expensive or difficult to get, and skill requirements to properly repair and maintain are difficult to find
- The existing equipment poses an unacceptable security risk, health and safety risk, or environmental risk and the cost to mitigate the risk exceeds the asset lifecycle replacement cost (BPA, 2014)

7

SPECIFIC AGING ASSET INTEGRITY MANAGEMENT PRACTICES

7.1   STRUCTURAL ASSETS

This chapter covers specific aspects of lifecycle management for structural infrastructure assets. Asset groupings include foundations, steel and re-enforced concrete structures, and pipe racks and overpasses. Chapter 5 described the importance of having good knowledge and information about the assets, upon which to develop a comprehensive lifecycle integrity management program. Each asset grouping sub section herein provides a listing of desired asset information, followed by a summary of warning signs of asset degradation. The chapter concludes with a listing of available standards and recommended practices for inspection and maintenance infrastructure from analogous industries.

7.1.1 Structure Foundations

This section addresses foundations for structures other than non-process buildings, including foundations for process equipment, process equipment structures, pipe racks and pipe bridges for example.

Asset Information. The following is a list of some typical information pertaining to industrial foundations:

Codes

- Company Standards/Practices
- American Concrete Institute (ACI)
- American Institute of Steel Construction (AISC-Seismic)
- Engineering contractor standards
- American Society of Testing and Materials (ASTM)

Design Information

- Soil Analysis and bearing capacity
- Piling system
- Concrete composition specifications and testing requirements
- Reinforcement specifications
- Specific foundation design loadings
- Foundation load type (static, dynamic, overturning)
- Seismic design criteria

Drawings

- Plot plans and layout
- Civil engineering
- Reinforcement details

Other

- Modification and repair history
- Changes in supporting structure use history

Aging Warning Signs. Some warning signs of aging structure foundations are provided that may prove useful:

- Foundation creep in tailings pond dykes with potential future breaching
- Cracked concrete at pipe rack and overpass supports due to structural creep from high loads and vibration
- Undermining of foundations due to changes in water table and sinkholes, etc.
- Undermining of foundations due to excessive dynamic loads
- Surface staining indicating rebar corrosion
- Spalling or dusting due to environmental attack
- Cracking due to frost heaving (e.g., cryogenic tank foundations)
- Sinking or distorted tank foundations (potential breach or failure at bottom seam)

There are many causes of foundation failure. Six main causes are listed below:

1. Soil type – especially expansive clay soil
2. Poorly compacted fill material
3. Slope failure, mass wasting
4. Erosion
5. Poor construction
6. Transpiration

Soil type – especially expansive clay soil. The most common kind of expansive clay can absorb so much water that it can swell by several hundred percent. The pressure from this degree of swelling can easily lift or "heave" most building foundations the size of residential homes. Soils expand with moisture and they contract with desiccation, causing up and down movements known as differential settlement. The structural integrity of the building can be maintained by providing underpinning for the foundation.

Poorly compacted fill material. If the fill material on a plot is not sufficiently compacted to support the weight of the structure above it, there will be foundation problems. The problem can be from the mix of odd fill materials, and from poorly compacted fill, or both.

Slope failure, mass wasting. Geologists use the term "mass wasting" to describe the movement of earth downhill. It could be "creep" which is slow, or "landslides" which are sudden. Slope failure as used here refers to "creep".

Underpinnings can act as a barrier to "creep", but the power of gravity is such that unless the underpinnings were specifically designed to stop slope failure, creep can still be an issue on sites exposed to slope failure.

Erosion. Erosion may be the most straightforward cause of settlement issues. It can come from poor drainage, uncontrolled water flow or lack of ground cover. If not identified early, erosion can wear away the soil around foundations, creating a new need for underpinning. The building in Figure 7.1-1 developed a wide crumbling crack in the lowest corner of the foundation. It sits on a hillside, and there is a water drainage area on the other side of the house resulting in erosion and creep.

Poor Construction. Many jurisdictions and communities now have building codes that require soil testing and engineer certification before and during the building process; consequently, poor construction is less likely the cause of foundation failure.

Figure 7.1-1. Image of a Building That Developed a Crumbling Crack

Transpiration. Transpiration is the process by which plants remove moisture from the soil. Trees withdrawing moisture from the soil in the summer can accelerate soil shrinkage in hot summer months. It is the expansion and shrinking or contraction of soils that disturb the foundation. It can exacerbate the problem with soil type described above.

Considerably activity is taking place under the soil, and it is often invisible from the surface. Thus, it is extremely important to detect any signs of foundation weakness early on by continuous monitoring and through inspections at routine intervals.

7.1.2 Support Structures

Support structures for equipment are addressed in this section. Support structures may be steel framed (see Figure 7.1-2) or of reinforced concrete construction. Fireproof insulation is often applied to the lower portion of steel structures to avoid weakening and collapse during a fire.

Equipment Supports Information. The following is a list of some typical information pertaining to industrial steel and reinforced concrete structure.

Codes and Standards

- American Institute of Steel Construction (AISC) Specification for Structural Steel Buildings
- American Society of Civil Engineers (ASCE) Minimum Design Loads for Buildings and Other Structures

Figure 7.1-2. Photo of Primitive Structural Supports

- International Code Council (ICC) International Building Code
- American Cement Institute (ACI)
- American Society of Civil Engineers (ASCE) Design of Reinforced Concrete
- American Society of Mechanical Engineers (ASME) [reinforces concrete slabs]
- American Petroleum Institute (API) Publication 2218 - Fireproofing Practices in Petroleum and Petrochemical Processing Plants
- API RP 583, Corrosion Under Insulation and Fireproofing (API, 2014)

Design Information

- Support list and installation date
- Equipment weights (including vessels full of water)
- Load capacity requirements
- Design calculations
- Seismic design criteria

Drawings

- Structure plan and elevation
- Floor layout
- Foundation details (if not covered elsewhere)

Other

- Modification and repair history

- Changes in supporting structure loads or use history

Equipment Supports Aging Warning Signs

Fireproofing

- Cracking resulting from damage or ingress of moisture behind fireproof layer and cyclic freezing and thawing. Mechanical impact or the force of a firewater jet can shatter the damaged fireproofing, leaving bare steel vulnerable to direct flame impingement and failure
- Spalled or cracked fireproofing allowing ingress of moisture and corrosion under fireproofing
- CUF may go undetected for many years. Figure 7.1-3 shows a sphere that collapsed as it was being filled with water for a pressure test. It is believed that CUF at the support legs was a contributing factor to the incident
- Environmental corrosion due to inherent process hazards

Steel Structures

- Environmental corrosion of steel structures from prior process emissions in older plants (e.g., carbon black and corrosive tail gas)
- Environmental corrosion of idled or out of service steel structures for extended periods

Figure 7.1-3. Liquefied Petroleum Gas (LPG) Storage Sphere Collapsed While Being Filled for a Hydrostatic Pressure Test

- Failed factory-applied coatings (blistering, pitting) that allow moisture to come in direct contact with the structure
- Corrosion at ground level from soil
- Loose or corroded platform grading (serious injuries to operators have occurred due to falling from elevated heights). Inspection during operator periodic rounds should be considered to identify issues.

Reinforced Concrete

- Deterioration appears as visual indications or discontinuities on exposed surfaces, including cracks, cracking patterns and width
- Crack distress such as efflorescence, rust stains and spalling
- Differential settling of surrounding structures

7.1.3   Piping Systems, Pipe Racks and Overpass Information

Pipe Racks and Piping Overpasses Information

Codes and Standards. Most of the same codes and standards listed for equipment supports apply to pipe racks, overpasses and bridges.

Design Information

- List of pipe racks and installation dates
- Piping weights (including full of water)
- Load capacity requirements
- Design calculations
- Seismic design criteria

Drawings

- Structural plan and elevation
- Foundation details (if not covered elsewhere)

Other

- Maintenance program records
- Modification and repair history
- Changes in supporting structure loads or use history

Pipe Racks and Piping Overpasses Warning Signs. Warning signs for support structures are applicable to piping supports. Additionally, for pipe racks and overpasses:

- Piping support hanger / bracket misalignment / damage
- Excessive piping movement (vibration / thrust) during operation.
- Lack of expansion / contraction to reduce stress on piping, valves and connections.

- Leakage from piping connections.
- Inadequate bolting
- Signs of vehicle impacts include gouging or dented steel, and damaged concrete
- Structural creep indicating high loads and/or vibration

7.1.4 Buildings

As a starter, the facility should compile an inventory listing of all the buildings to be covered by the inspection program. The inventory of the buildings should summarize useful information such as building type, location, year built, use, occupancy, electrical classification, ventilation, and fire protection systems. The list should also identify what department is responsible for the upkeep of the building.

The next step is to prioritize the list in terms of safety and operational criticality based on potential risk. The purpose of the criticality ranking is to help allocate resources where they can have the most benefit in reducing risk. If asset information is misplaced or lacking (not unusual), it will identify which assets should be addressed first for compiling documentation. Figures 7.1-4 through 7.1-6 show the effects on aging on several buildings. The following is a list of some typical information pertaining to industrial buildings:

Figure 7.1-4. Chemical Plant Shelter Showing Signs of Severe Deterioration

Figure 7.1-5. Building Presenting Aging Signs

Figure 7.1-6. Photo of Aged Chemical Silos

Building Information

Codes

- International Building Code (IBC)
- International Mechanical Code (IMC)
- National Electric Code (NEC)
- National Fire Protection Association (NFPA) Standards
- American Society of Heating, Refrigerating, and Air-Conditioning Engineers (ASHRAE) Standards
- Air Movement and Control Association (AMCA) Standards
- Other standards may apply depending on jurisdiction

Design Information

- Structural Loads (equipment, roof, wind, onside shock)
- Ventilation rates
- Sprinkler water application rates
- Maximum occupancy

Drawings

- Civil and structural
- Floor layout
- Electrical classification
- Fire protection systems
- Electrical wiring
- HVAC systems
- Gas, smoke and fire detection

Other

- Modification and repair history
- Changes in use history
- Fire or damage history

A brief description of any schedule or periodic maintenance performed or other activities that monitor the conditions should be included with a pointer to any formalized maintenance/inspection program.

Building Aging Warning Signs. Some warning signs of building aging are provided that may prove useful. Firefighting organizations are keenly aware of warning signs of structurally unstable buildings, because their lives may be in danger. The following are some warning signs of severe building aging:

- deterioration of mortar joints and masonry
- cracks
- signs of building repair

- large open spans
- bulging and bowing walls
- sagging floors, ceilings, and beams
- soft or spongy footing
- noticeable creaking sounds
- poor condition of engineered lumber, truss joints and nail plates

The indicators of building movement can be detected by monitoring the openings in the walls, the doors, windows, and passageways. Are they square? Do the doors still fit? Can you close them? Have the windows cracked for no reason? Are the frames square in the window frame? If not, the building is probably moving.

The partial collapse of a Harlem, New York apartment house killing three people was the result of prolonged water damage. Former tenants in the building revealed that the basement had been flooded for months. Water seepage from the wall that eventually failed, a major crack in that wall, badly sloping floors and gaps between floors and the walls that held them up were all contributing factors. The water came from a large pipe that entered the building through the same wall that collapsed and allowed water seepage from that wall. People in the building industry suggested that water is often a culprit in structural damage to older buildings, because it erodes mortar between stones and the earth beneath walls. (New York Times, 1995).

7.1.5　Inspection and Maintenance RAGAGEPs

Maintenance and Inspection RAGAGEP. Chapter 6 addressed the development and implementation of a lifecycle integrity management program for aging assets through the use of maintenance and inspection practices. While there are many recommended practices and guidelines for inspection of traditional process equipment, those targeted at infrastructure assets are few in number and from organizations less associated with the process industries. However, there are some recommended practices and guidelines from other industries such as electric utility, nuclear, and transportation that provide analogies for chemical plant infrastructure, as well as those from professional organizations like API and ASCE. Table 7.1-1 provides a listing of types of infrastructure and analogous RAGAGEP for inspections.

Table 7.1-1. Analogous Inspection Practices for Structures

Infrastructure Type	Analogous Industry	Practices and Guidelines

Outside Battery Limits (OSBL) Interconnecting Piping	Petroleum Refining	API 570, *Piping Inspection Code: Inspection, Repair, Alteration, and Rerating of In-service Piping Systems* API RP 574 *Inspection Practices for Piping System Components* API RP583 *Corrosion Under Insulation and Fireproofing* API Recommended Practice 580, *Risk-Based Inspection*
Steel Support Structures	Power Transmission	ASCE Conference, *Electrical Transmission and Substation Structures 2015: Best Practices for Transmission Line Inspections and Recommended Inspection Techniques* NACE SP0415-2015/IEEE Std. 1895, *"NACE/IEEE Joint Standard Practice for Below-Grade Inspection and Assessment of Corrosion on Steel Transmission, Distribution, and Substation Structures"* NACE SP0315-2015/IEEE Std. 1835, *"NACE/IEEE Joint Standard Practice for Atmospheric (Above Grade) Corrosion Control of Existing Electric Transmission, Distribution, and Substation Structures by Coating Systems"*
Foundations	Power Transmission	ASCE, Electrical Transmission and Substation Structures 2015: *Deep Foundations - Combining Construction Methods, Engineering, Inspection, and Testing*
Pipe Bridges & Trestles	Highway Transportation	American Association of State Highway Transportation Officials (AASHTO), *Manual for Bridge Element Inspection, 1st Edition, with 2015 Interim Revisions*

Table 7.1-1. Analogous Inspection Practices for Structures, continued

Infrastructure Type	Analogous Industry	Practices and Guidelines
Reinforced Concrete Pipe	Transportation, Civil Engineering	Federal Highway Administration (FHWA), *Culvert Inspection Manual* ASCE Conference Pipelines 2009: Infrastructure's Hidden Assets: *Unmanned RFTC Inspection of Large Diameter Pipe*

Inspection Checklist. As indicated in Chapter 6, visual inspection of assets is a viable method at least for a preliminary inspection of above ground systems. The objective of a preliminary inspection is to obtain initial analytical information to assess the physical adequacy of an existing asset. This inspection is usually a walking examination of the asset to visually evaluate the structural components.

Table 7.1-2 provides some examples of signs of aging infrastructure to look for when performing a visual inspection.

If no inspection records are available, a list of structures to be inspected should be prioritized based on the risk as discussed in Chapter 4 and age of the structure. A schedule should be set up for initial inspection.

Table 7.1-2. Example Checklist for Structural Assets

A - BUILDING & CIVIL STRUCTURES	Y	N	NA	Remarks
1. Are there wear patterns on surface, areas of missing protective coating exhibiting feathered edges of the sound coating?	☐	☐	☐	
2. Are there cracking, fractures or exposed aggregate on walls, beams, or supporting concrete structures?	☐	☐	☐	
3. Is there exposed concrete reinforcement or rust stains on support structure?	☐	☐	☐	
4. Are there cracking or distortion of elastomeric sealants, caulking, or gaskets?	☐	☐	☐	
5. Is there misalignment at construction joints or pipe joints, misaligned doors?	☐	☐	☐	
B - STEEL SUPPORT STRUCTURE	Y	N	NA	Remarks
6. Is there corrosion attack and accumulation of corrosion products on surface?	☐	☐	☐	
7. Are there cracks or blisters in protective coating?	☐	☐	☐	
8. Are there loose anchor bolts?	☐	☐	☐	
9. Are there corroded bolts?	☐	☐	☐	
10. Is there blistering or spalling of protective coatings, including fireproofing?	☐	☐	☐	
C - CIVIL FOUNDATIONS	Y	N	NA	Remarks
11. Are there signs of settling, cracks, out of level?	☐	☐	☐	
12. Has soil eroded or sunk near foundation?	☐	☐	☐	
13. Are there signs of upheaving from freeze/thaw cycle, cracks, out of level?	☐	☐	☐	

7.2   ELECTRICAL DISTRIBUTION AND CONTROLS

This next section addresses standards and practices that support the aging infrastructure principals of electrical distribution systems and controls.

7.2.1 Electrical System

Electrical System Information. The following is a list of some typical information pertaining to industrial electrical power systems that should be compiled.

Codes and Standards

- National Fire Protection Association(NFPA)
- National Electric Code (NEC-70)
- Institute of Electrical and Electronics Engineers (IEEE) 3000 Standards Collection for Industrial and Commercial Power Systems
- National Electrical Contractors Association (NECA)
- National Electrical Installation Standard (NEIS)
- InterNational Electrical Testing Association (NETA) MST-2001

Design Information

- Equipment list with installation date, manufacture and ratings
- Component and wire specifications
- System loads/electrical demand by load center
- Emergency generator specification and loads
- Uninterruptible Power Supply (UPS) loads

Drawings

- Electrical One Lines
- Underground cable routing
- Emergency generator
- Transformers
- Area electrical classification

Other

- Manufacturer's maintenance manuals
- Maintenance program records
- Modification and repair history
- Changes in electrical loads or use history

Electrical System Inspection. Examples of inspection practices for electrical infrastructure are listed in Table 7.2-1. The use of checklists during maintenance and inspection activities is a good practice to ensure that inspections are thorough. Table 7.2-2 is a sample checklist for electrical infrastructure.

Table 7.2-1. Inspection Practices for Electrical Infrastructure

Infrastructure Type	Analogous Industry	Practices and Guidelines
Electrical Equipment	Power and Utility	NFPA *70B, Recommended Practice for Electrical Equipment Maintenance.*
		FM *Global's Loss Prevention Data Sheets*
		InterNational Electrical Testing Association (NETA) MTS-2001, *Maintenance Testing Specification for Electrical Power Distribution Equipment and Systems.*

		Infraspection Institute, _Standard for Infrared Inspection of Electrical Systems and Rotating Equipment._
		ASTM, _E 1934 Standard Guide for Examining Electrical and Mechanical Equipment with Infrared Thermography._
		Danish Technology Institute, _TTCTRAN.015 Guideline for Thermographic Inspection in Electrical Installations._

Table 7.2-2. Example Checklist for Electrical Systems

ELECTRICAL - CABLES & CONDUITS	Y	N	N A	Remarks
1. Does electric heat tracing have cracks in insulation or exposed wires?	☐	☐	☐	
2. Are there broken or tightly bent flexible conduit, broken or damaged rigid conduit, loose or damaged conduit connectors?	☐	☐	☐	
3. Are there missing or damaged terminations or pull box/fixture covers, loose or damaged cable tray sections, loose, corroded or damaged environmental seals?	☐	☐	☐	
4. Are there missing padding, fairleader, bells, or bushings from cable drop-outs in cable tray or from cable entry points into conduit?	☐	☐	☐	
5. Are there elevated ambient temperature and humidity in area occupied by cable and other electrical equipment?	☐	☐	☐	
6. Are there corroded fasteners in electrical manhole, water marks on manhole walls?	☐	☐	☐	
ELECTRICAL - MOTORS	Y	N	N A	Remarks
7. Is there disturbed or loose foundation bolts and grout, cracked welds at the housing feet, vibration felt when casing is touched?	☐	☐	☐	
8. Is there grease leakage from motor bearing housing?	☐	☐	☐	
9. Is there high-pitched noise?	☐	☐	☐	
10. Is there burning smell or signs of burnt insulation?	☐	☐	☐	
11. Are there signs of change in color of paint on the casing?	☐	☐	☐	

Table 7.2-2. Example Checklist for Electrical Systems, continued

ELECTRICAL - MOTORS	Y	N	N A	Remarks
ELECTRICAL - BREAKERS AND SWITCHES	Y	N	N A	Remarks
12. Is the face of breaker warm or hot to touch?	☐	☐	☐	

13. Is there buzzing or crackling sound when switching?	☐	☐	☐	
14. Are there signs of corrosion on electrical terminals?	☐	☐	☐	
15. Is there burning smell or signs of burnt insulation?	☐	☐	☐	
16. Are there signs of discolored/cracked conductor insulation?	☐	☐	☐	
17. Are all indicator lights functional?	☐	☐	☐	
18. Are electrical rooms / areas clean, cool, tight, dry and free of storage?	☐	☐	☐	
19. Is there combustible storage beyond 5 feet (1.5 meters) around electrical equipment?	☐	☐	☐	
20. Is there evidence of past (or current) water leakage into the room or structure?	☐	☐	☐	
21. Are electrical rooms climate controlled, where required, based on the operating environment?	☐	☐	☐	
22. Are electrical rooms under a positive pressure to prevent contaminates from entering the room (where required)?	☐	☐	☐	
23. Is the outdoor electrical enclosure / building water and contaminate tight?	☐	☐	☐	
24. Is there evidence of excessive corrosion in the electrical equipment enclosure or past water leakage?	☐	☐	☐	
ELECTRICAL - GROUNDING & BONDING	**Y**	**N**	**N A**	**Remarks**
25. Are any connections loose?	☐	☐	☐	
26. Is there visible corrosion on connections?	☐	☐	☐	
27. Is there visible damage to the wire?				
BATTERY SYSTEMS	**Y**	**N**	**N A**	**Remarks**
28. Are battery rooms clean, cool, tight, dry and free of storage?	☐	☐	☐	
29. Are battery terminal connections not corroded?	☐	☐	☐	
30. Is the battery support rack not corroded?	☐	☐	☐	
31. Is the battery charger operating?	☐	☐	☐	

Thermal scanning is one of the common inspection techniques for electrical components, in addition to other solid dielectric and functional tests to verify the integrity of the electrical equipment. This technique identifies electrical components overheating due to high resistance or high current flow.

Overheating of the electrical components might be due to lose connections, overloaded circuits or phase imbalance.

Contractors or trained internal electrical technicians typically perform annual thermal scans with specialized equipment. Typically, thermal scans are completed every one to three years, depending on equipment conditions. In order to perform infrared on electrical switchgear that cannot be opened for inspection, specially designed view ports can be installed in the electrical cabinets.

Figures 7.2-1 through 7.2-6 show some images of thermal scans indicating issues that need to be corrected.

Figure 7.2-1. Motor Control Center (MCC) Thermal Scan of a Phenol Unit –
Photo 1

Figure 7.2-2. Motor Control Center (MCC) Thermal Scan of a Phenol Unit - Photo 2

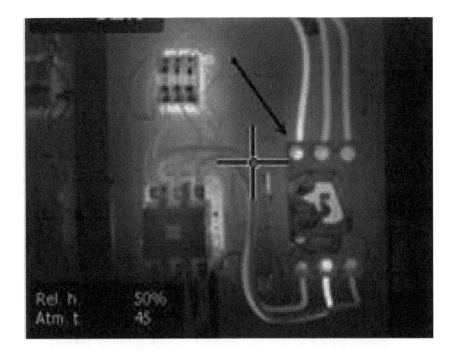

Figure 7.2-3. Thermography Image Showing Hot Terminals

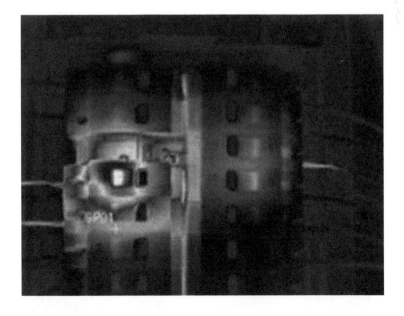

Figure 7.2-4. Damaged Contacts in Lighting Panel Circuit Breaker

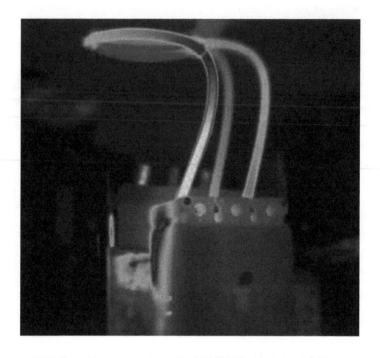

Figure 7.2-5. Loose A-Phase on 3 Phase Circuit Breaker and Possible Unbalanced
Load

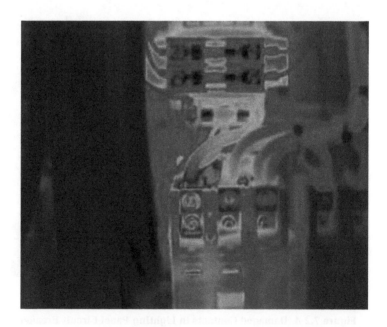

Figure 7.2-6. Loose Connection on A-Phase in a 2 Speed Motor Contactor

Transformer testing to verify integrity can include dissolved gas analysis on the transformer oil to test for various gases and conditions which are indicators of internal damage mechanisms, including arcing and overheating. This in in addition to other solid dielectric testing which can include power factor and insulation resistance testing.

Checking resistance of grounding systems and continuity of bonding systems are two common practices, in addition to visual inspection, to monitor performance.

Refer to standards referenced in Table 7.2-1 for additional recommended inspection practices.

Electrical System Aging Warning Signs. Often there are warning signs that precede a potential catastrophic failure. Some examples for electrical infrastructure are listed below.

- Transformer tank vibration indicating core and winding looseness, deterioration of the pressboard due to moisture or heat may cause shrinkage and looseness. In order for a transformer to withstand through faults or switching surges that include heavy load conditions, the core and windings must be securely blocked and clamped
- Sound reaching the transformer tank wall indicating Partial Discharges (PD). Insulation breakdown from PD can reach a point where it threatens the life of a transformer
- Frayed or cracked wire insulation
- Overheating of transformers, motor control centers, and breakers
- Non-sealed electrical boxes in Class II (combustible dust) environments
- Frequent arcing observed in darkness or during rainstorms indicating deterioration (e.g., ceramic bushings) of electrical components on elevated power lines

Prior to the 1950's electrical wiring was insulated with hard rubber or non-conducting cloth wrap. Some older electrical wiring within commercial buildings (including chemical plants and warehouses) utilized a system of knobs and tubes which separated hot leads from combustible materials such as wood. This technology has long since been outdated. Codes now require conduit or PVC insulated cable. These are less sensitive to heat and moisture.

Vintage wiring within older buildings is subject to electrical shorts leading to possible power failures or fires. As older insulating materials age, they tend to dry out or split and they may lose their insulating properties. The obvious solution may appear to be a system retrofit or replacement. However, this can introduce additional problems such as exposure of mold or asbestos within walls. A holistic and integrated approach to retrofitting building electrical systems must be considered. Consider all the options and don't be caught by surprise.

Figure 7.2-7 shows some medium voltage (15kV) indoor open-air switchgear in a 1960s-era electrical substation. Note that the front wooden barriers are only 5 feet high making it possible for personnel to come into contact with energized components. The side barriers are not fully enclosed and do not reach the top of the compartments. While there are no safety regulations that dictate layout requirements in the electrical substations, current practices preclude open air technology making the facility pictured below obsolete.

Figure 7.2-7. Medium Voltage (15kV) Indoor Open-Air Switchgear

Medium voltage switchgear is currently classified as metal-enclosed, metal clad, arc-resistant, or GIS (gas insulated). This older open-air technology is still encountered in rural outdoor installations, but is completely obsolete in modern substations. Any changes or modifications to the facilities in the photo would likely trigger a major upgrade.

7.2.2   Control System

The primary issue with process control systems is functional aging or obsolescence. The equipment becomes difficult to maintain due to scarcity of spare parts, and subject to frequent unavailability. The infrastructure that supports the process control system such as uninterruptible power supply, instrument air, and emergency power may also encounter aging issues.

Control System Infrastructure Information.

Codes, Standards and Practices

- NFPA 72 National Fire Alarm and Signaling Code (UPS)
- NFPA 111 Storage of Electrical Energy

- American National Standards Institute (ANSI) / Instrument Society of America (ISA), ANSI/ISA 7 Quality Standard for Instrument Air
- NFPA 110 Emergency and Standby Power Systems
- CCPS Safe Automation Guidelines (2016)

Design Information

- Sized equipment list and installation dates
- Load design basis
- Connected loads
- Instrument air design conditions (pressure, dew point, lube oil)
- I/A equipment specifications and redundancy
- UPS capacity basis and connected loads

Drawings

- Piping and Instrumentation Diagram (P&ID)
- UPS electrical one line
- Emergency Power P&ID

Other

- Manufacturer's maintenance manuals
- Statutory (state mandated) air receiver inspection reports
- Maintenance and testing practices
- Modifications and repair history
- Incident history

Control Systems Infrastructure Inspection. Examples of inspection practices for control systems infrastructure are listed in Table 7.2-3.

Table 7.2-3. Inspection Practices for Control Systems Infrastructure

Infrastructure Type	Analogous Industry	Practices and Guidelines
Control System Infrastructure	Nuclear, Environmental, Military, Energy	US Army Technical Manual TM 5-697, *Commissioning of Mechanical Systems for Command, Control, Communications, Computer, Intelligence, Surveillance, and Reconnaissance (C4isr) Facilities*, Headquarters, Department of The Army, 2006
		NUREG/CR-5419 *Aging Assessment of Instrument Air Systems in Nuclear Power Plants*, Brookhaven National Laboratory, 1990
		QA Handbook Vol II, Section 11.0, *Instrument Equipment Testing, Inspection and Maintenance*, 2008
		DOE, *Office of Industrial Technology, Maintenance of Compressed Air Systems for Peak Performance*, Fact Sheet #5, 1998

		Maintenance for UPS Systems, http://electrical-engineering-portal.com/maintenance-for-ups-systems, 2010

Control Systems Infrastructure Warning Signs

Instrument Air Facilities

- Inadequate performance during power outages
- Inadequate performance during cold weather (wet I/A)
- Inadequate performance of pneumatic components (low I/A oil content)
- Increase in I/A failure incidents

UPS Facilities

Some disturbance causes of UPS, which can result in system shutdown, damage to computers and electronic boards, accelerated aging or stress breakdown of components, are shown on Table 7.2-4.

Instrumentation

For monitoring instrumentation, Environmental Protection Agency (EPA) estimates a seven (7) year lifespan for most monitoring instruments and a somewhat longer lifespan for more permanent types of equipment (instrument racks, monitoring shelters etc.). Organizations involved in monitoring may be able to prolong the life of equipment but in doing so they may run the risk of additional downtime, more upkeep and a greater chance of data invalidation, while losing out on newer technologies, better sensitivity/stability and the opportunities for better information management technologies (EPA, 2008).

Table 7.2-4. List of UPS Disturbance Causes

Disturbance	Characteristic	Main Cause
HF Transients	Sudden, major and very short jump in voltage.	Similar to a voltage spike. Atmospheric phenomena (lightning), static discharges and switching.
Short Duration	< 1 μs Amplitude < 1 to 2 kV at frequencies of several tens of MHz.	Starting of small inductive loads, repeated opening and closing of low-voltage relays and contactors.
Medium Duration	1 μs and ≤ 100 μs Peak value 8 to 10 times higher than the rated value up to several MHz.	Faults (lightning) or high-voltage switching transmitted to the low voltage by electromagnetic coupling.
Long Duration	> 100 μs	Stopping of inductive loads or high-voltage faults transmitted to the

	Peak value 5 to 6 times higher than the rated value up to several hundred MHz.	low-voltage system by electromagnetic coupling.

Source: Schneider Electric (Schneider, 2012)

7.3   EARTHWORKS: ROADS, IMPOUNDMENTS, AND RAILWAYS

7.3.1   Roads

Road Construction Information. The following is a list of some typical information pertaining to roads systems that should be compiled.

Codes and Standards

Road ways are typically the responsibility of the owner companies, including construction practices, maintenance, and repair. There is a large body of standard specifications for road construction, including USA State Departments of Transportation (SDOT). Some examples are provided below.

- Standard Specifications for Roads & Bridges on Federal Highway Projects, United States Department of Transportation, Federal Highway Administration, FP-14, 2014
- Various Standard Specifications (e.g., aggregate, asphalt mixtures, cement), American Association of State Highway & Transportation Officials (AASHTO)
- Various Standard Specifications (e.g., aggregate, analysis of asphalt mixtures, asphalt), American Society of Testing and Material (ASTM)
- Standard Specifications for Construction of Highways, Streets & Bridges, Texas Departments of Transportation, 2014
- Standard Specifications, State of California, California State Transportation Agency, Department of Transportation, 2015
- Highway Design Manual, Cal DOT, 6th Edition 2006 plus revision changes

Design Information

- Aggregate specification
- Asphalt or Concrete specification
- Subsurface preparation
- Load weight limit
- Drainage design details

Drawings

- Plot plan showing road layout
- Road construction detail cross section
- Storm drainage system

Other

- Modification and resurfacing/repair history
- Accident history

Road Inspection. Maintenance inspection guidelines are available from many state transportation departments and agencies. Some international examples of inspection practices for road infrastructure are listed in Table 7.3-1.

The use of checklists during maintenance and inspection activities is a good practice to ensure that inspections are thorough. Table 7.3-2 is a sample checklist for Road infrastructure.

Although there are no specific inspection frequency requirements, the quality of the roads should be checked annually. Spring inspection is most advantageous time to identify issues created by ground freezing or thawing.

Table 7.3-1. Inspection Practices for Road Infrastructure

Infrastructure Type	Analogous Industry	Practices and Guidelines
Roads & Highways	Federal and State Agencies	*Road Safety Inspections*, Road infrastructure safety protection – core-research and development, Increasing safety and reliability for surface transport (RIPCORD-iSEREST), Work package 5, EU, 2005. *Road Safety Inspections Guidelines and Checklists*, Tallinn University of Technology, Lithuanian Road Administration, European Regional Development Fund, 2012.

Table 7.3-2. Example Checklist for Roads Maintenance and Inspection

Roads - Flexible Pavement	Y	N	NA	Remarks
1. Are there longitudinal or transverse cracks > 0.5" wide?	☐	☐	☐	
2. Is there alligator cracking > 10% of surface on both wheel paths?	☐	☐	☐	
3. Are wheel ruts deeper that 0.5 inches?	☐	☐	☐	
4. Is there block cracking >20%?	☐	☐	☐	
5. Are there noticeable potholes?	☐	☐	☐	
6. Is there surface weathering and reveling?	☐	☐	☐	
7. Is there deteriorated edge cracking?	☐	☐	☐	

	Y	N	NA	
8. Is there significant patching?	☐	☐	☐	
9. Are there areas that do not drain properly?	☐	☐	☐	

Table 7.3-2. Example Checklist for Roads Maintenance and Inspection, continued

Roads - Concrete	Y	N	NA	Remarks
10. Is there spalling within 2" of longitudinal or transverse joints?	☐	☐	☐	
11. Is there longitudinal cracking > 50 ft. per 1000 sq. ft. of pavement?	☐	☐	☐	
12. Is there deteriorated slab patching (faulting or settling)?	☐	☐	☐	
13. Is there D-cracking (caused by freeze/thaw expansion)?	☐	☐	☐	
14. Is there faulting (uplift) at joints or cracks?	☐	☐	☐	
15. Are there shattered slabs (intersecting cracks, caused by overloading or inadequate support, that divide the pavement into four or more pieces)	☐	☐	☐	
Roads - Concrete	Y	N	NA	Remarks
16. Is there joint sealant damage at transverse joints (including extrusion, hardening, adhesive failure (loss of bond), cohesive failure (splitting), or complete loss of the sealant. The presence of weed growth in the joint is also an indication of joint seal damage)?	☐	☐	☐	
17. Are there areas that do not drain properly?	☐	☐	☐	

Road Aging Warning Signs. Often there are warning signs that precede a potential catastrophic failure. Some examples for road infrastructure warning signs are; degradation of roadway surface, ruts, cracking and faulting of concrete and asphalt from exposure to extremes in temperature, chemical spills, and heavy truck traffic.

7.3.2 Earthworks Infrastructure: Trenches, Dikes and Storage Ponds

Earthworks Infrastructure: Information. The following is a list of some typical information pertaining to earthworks systems that should be compiled.

Codes and Standards

- American Petroleum Institute (API), Standard 2610 Design, Construction, Operation, Maintenance, and Inspection of Terminal and Tank Facilities (for dikes and berms)

- California Stormwater Quality Association, Stormwater Best Management Practice Handbook: Construction
- United States Department of Agriculture (USDA), Natural Resources Conservation Service, Agriculture Handbook Number 590, Ponds — Planning, Design, Construction
- Design of Urban Stormwater Controls, Water Environment Federation (WEF) and the American Society of Civil Engineers (ASCE), Manuals of Practice (MOP) MOP 87
- State environmental regulations
- Company or engineering contractor standards and practices

Design Information

- Pond capacity basis
- Detention depth and freeboard
- Storage time
- Pond sealing method (clay, liner, etc.)
- Embankment design and stabilization
- Dike size
- Dike construction
- Dike coating

Drawings

- Plan and layout
- Berm cross section detail
- Pipe penetrations
- Intake detail

Other

- Modification and repair history (e.g., work orders, management of change files)
- Capacity requirement changes (documented and undocumented)

Earthworks Infrastructure: Inspection. Several state agencies have safety inspection requirements for earthwork infrastructures. Examples of inspection practices for earthworks infrastructure are listed in Table 7.3-3.

The use of checklists during maintenance and inspection activities is a good practice to ensure that inspections are thorough. Table 7.3-4 is a sample checklist for earthworks infrastructure.

Table 7.3-3. Inspection Practices for Earthwork Infrastructure

Infrastructure Type	Analogous Industry	Practices and Guidelines

Earthen Embankments	Dam Safety	Inspection of Embankment Dams, Association of State Dam Safety, www.damsafety.org (includes warning signs) Inspection Checklist Guidelines, Iowa Department of Natural Resources, www.iowadnr.gov/
Detention Ponds	Storm Water Management	Stormwater Operation& Maintenance, A Resource Guide, Boise Public Works, Boise City, 1999 Stormwater Facility Inspection and Maintenance Handbook, Whatcom County, WA, 2011
Railroad Tracks	Transportation	Unified Facilities Criteria (UFC) Railroad Track Maintenance & Safety Standards, UFC 4-860-03, 2008

Table 7.3-4. Example Checklist for Earthworks Infrastructure Maintenance and Inspection

Earthen Dams and Dikes	Y	N	NA	Remarks
1. Are there signs of embankment instability, e.g., sloughing, sliding, cracking or bulging, or unusual settling?	☐	☐	☐	
2. Are there embankment slumps or slides? This instability may be caused by excessive seepage, poor compaction, foundation problems or too steep side slope.	☐	☐	☐	
3. Is there embankment longitudinal or transvers cracking? Deep longitudinal cracks (parallel to the length of the embankment) are usually a precursor to embankment slumps or slides. Transverse cracking (perpendicular to the length of the embankment) is typically caused by poor compaction effort during construction.	☐	☐	☐	
4. Are there signs of burrowing animals?	☐	☐	☐	
5. Are there embankment rills and gullies? Pay particular attention to the groin areas.	☐	☐	☐	
6. Are there sinkholes? Sinkholes are likely a sign of internal erosion within the embankment, typically along a conduit.	☐	☐	☐	
7. Is there uncontrolled seepage? All embankment dams will have some seepage. Springs, boils or simply wet areas with or without wetland vegetation are an indication of uncontrolled seepage.	☐	☐	☐	
8. Is there woody vegetation growth? The embankment and groins should be kept free of all woody growth. The tree and brush free zone should extend at least 20 feet beyond the embankment toe and groins.	☐	☐	☐	
Detention - Tailings Ponds / Dams	Y	N	NA	Remarks
9. Is there slumping or sloughing of walls? If over 4" of slumping, consult with an engineer.	☐	☐	☐	
10. Is there any erosion or scouring especially at inlets and outlets? If leaks or soft spots are found, consult with an engineer.	☐	☐	☐	
11. When pond level permits, are there visible holes in the liner?	☐	☐	☐	
12. Is there evidence of burrowing animals or insects?	☐	☐	☐	
13. Is integrity of access ramp stable and clear for heavy equipment?	☐	☐	☐	
14. Are there trees and shrubs that shade sidewall grass or that might have problem roots near pipes and structures?	☐	☐	☐	
15. Is the energy dissipating rip-rap adequate?	☐	☐	☐	

Table 7.3-4. Example Checklist for Earthworks Infrastructure Maintenance and Inspection, continued

Concrete Dikes	Y	N	NA	Remarks
16. Are there cracks in the walls or floor of the dike?	☐	☐	☐	
17. Is there sealant cracking in the wall penetrations?	☐	☐	☐	
Concrete Dikes	Y	N	NA	Remarks
18. Is there deteriorated slab patching (faulting or settling)?	☐	☐	☐	
19. Is there D-cracking (caused by freeze/thaw expansion)?	☐	☐	☐	
20. Is there faulting (uplift) at joints or cracks?	☐	☐	☐	
21. Is there evidence of sealant erosion?	☐	☐	☐	
22. Is there joint sealant damage at transverse joints (including extrusion, hardening, adhesive failure (loss of bond), cohesive failure (splitting), or complete loss of the sealant. The presence of weed growth in the joint is also an indication of joint seal damage)?	☐	☐	☐	
23. Are there areas that do not drain properly?	☐	☐	☐	

Typically, inspection frequencies for dikes, ponds and trenches are defined by regulations. Some examples of these are:

- US EPA Resource Conservation and Recovery Act
- US EPA Spill Prevention, Control, and Countermeasure

Earthworks Aging Warning Signs

- Foundation creep in storage pond earth dams leading to future breach and flooding
- Sealing failure in waste disposal ponds and chemical tailings ponds allowing leaching of hazardous materials into subsoil
- Deteriorated surface coating, erosion or worn down embankment of tank farm earthen dikes, with potential for overtopping resulting in a loss of containment event

7.3.3 Railways and Spurs

Railroad Track Information. The following is a list of some typical asset management information pertaining to rail spurs.

Codes and Standards

Rail way spurs are typically the responsibility of the railroad companies, including construction practices, maintenance, and repair.

Design Information
Information may be available from the railroad company, if required.

Drawings

- Plot plan showing rail spur locations
- Spill containment at loading/unloading spot

Other

- Railroad inspector's reports
- Modification and repair history
- Accident history

7.3.3.2 *Railroad Track Maintenance Inspection*

The use of checklists during maintenance and inspection activities is a good practice to ensure that inspections are thorough. Table 7.3-5 is a sample checklist for rail spur infrastructure.

Table 7.3-5. Example Checklist of Inspection Practices for Rail Spur Infrastructure

Rail - Roadway	Y	N	N A	Remarks
1. Is there evidence of ballast subgrade attrition (mud pumping into the ballast)?	☐	☐	☐	
2. Is there erosion of embankments and cut slopes?	☐	☐	☐	
3. Is there embankment sliding and slippage?	☐	☐	☐	
4. Is there settling at approaches to bridges and roads crossings?	☐	☐	☐	
5. Are there washout under or adjacent to tracks?	☐	☐	☐	
6. Are there potential slope stability problems?	☐	☐	☐	
Rail - Drainage	**Y**	**N**	**N A**	**Remarks**
7. Is brush restricting drainage?	☐	☐	☐	
8. Has soil drift restricted drainage?	☐	☐	☐	
9. Is excessive ice and snow restricting proper drainage?	☐	☐	☐	
10. Are there other obstructions interfering with the flow of water?	☐	☐	☐	
11. Is drainage an issue at turnouts, road crossings, bridge ends?	☐	☐	☐	
Rail - Ballast	**Y**	**N**	**N A**	**Remarks**
12. Is ballast clean and free of vegetation?				

13. Is ballast full crib and not overtopping the ties, and not interfering with turnout switches?	☐	☐	☐	

Table 7.3-5. Example Checklist of Inspection Practices for Rail Spur Infrastructure, continued

Rail - Ties	Y	N	N A	Remarks
14. Are ties broken through?	☐	☐	☐	
15. Are ties split or impaired to the extent they will not hold spikes?	☐	☐	☐	
Rail - Ties	**Y**	**N**	**N A**	**Remarks**
16. Is the tie deterioration such that the tie plates can move >0.5" laterally?	☐	☐	☐	
17. Is the tie deterioration such that the tie is cut by the tie plates >2"?	☐	☐	☐	
18. Is tie cut by wheel flanges, dragging equipment, or fire to a depth of 2" within 12" of the base of the rail or load bearing area?	☐	☐	☐	
19. Are rail joint bar bolts loose or missing?	☐	☐	☐	
20. Is tie rotted, hollow, or generally deteriorated to the point that a substantial amount of material is decayed or missing?	☐	☐	☐	

Railways and Spurs Warning Signs. Railway warning signs of possible derailment if not heeded and corrected.

- Deterioration of railway lines and spurs – wooden railways ties split and cracked, ignored and subjected to occasional loads and possible derailment
- Railway spurs seasonally under water and redistributing of ballast due to subgrade attrition
- Loose or sunken tie plates, missing or loose spikes, and cracked or broken joint bars
- Loose or missing bolts at joint bars
- Impact damage at rail spurs and loading racks (detached walkways and damaged loading arms, cables, etc.) – often not reported by third party contractors
- Spur not properly inspected due to confusion over ownership

Figure 7.3-1. Picture of Vintage Tank Car Fastened with Rivets

(how do you inspect this and who is qualified to do so?)

7.4   MARINE FACILITIES: TERMINALS AND JETTIES

This section addresses standards and practices specific to Marine Facilities that support following aging infrastructure principles.

7.4.1   Marine Facilities Information

The following is a list of some typical information pertaining to Marine Facilities that should be compiled.

Codes and Standards. Marine facilities are regulated by the US Coast Guard and State agencies such as California State Lands Commission (CSLC). The responsibility of the owner companies includes compliance with construction standards, and maintenance, and inspection. Some example standards and practices are provided below.

- *Marine Oil Terminal Engineering & Maintenance Standards* (MOTEMS), California State Lands Commission (CSLC), Marine Facilities Div. 2013
- *Guidelines for Marine Oil & Petrochemical Terminal Design*, World Association of Waterborne Transport Infrastructure (PIANC, 2014)
- *Handling Dangerous Cargos at Waterfront Facilities*, DHS, US Coast Guard Regulations 33CFR-126 (plus parts 127 [LNG/LHG], 154 & 156 [Oil & HAZMAT])

Design Information

- Terminal operating limits. The physical boundaries of the facility defined by the berthing system operating limits, along with the vessel size limits and environmental conditions
- Mechanical and electrical equipment specifications (e.g., loading arms, cranes and lifting equipment including cables, piping/manifolds and supports, oil transfer hoses, fire detection and suppression systems, vapor control system, sumps/sump tanks, vent systems, pumps and pump systems, communications equipment, electrical switches and junction boxes, emergency power equipment)
- Mooring and berthing analyses. Analyses consistent with the terminal operating limits and the structural configuration of the wharf and/or dolphins and associated hardware
- Structural and seismic analyses and calculations
- Piling specifications
- Description of Emergency Shutdown (ESD) and isolation systems

Drawings

- Plot plan showing jetty and trestle layout
- Mooring and breasting dolphin configuration
- Detailed foundation and structural drawing
- Loading arm mechanical drawings
- Fire protection system
- P&IDs and piping arrangement
- Electrical one lines

Other

- Background data on the terminal - description of the service environment (wind, waves, currents), extent and type of marine growth
- Inspection/testing data
- Geotechnical report
- Terminal fire protection plan
- Pipeline stress and displacement analyses
- Accident history
- Spill control and emergency response procedures

7.4.2 Marine Facility Inspection

Maintenance inspection guidelines are available from various transportation departments and agencies. Some international examples of inspection practices for marine infrastructure are listed in Table 7.4-1.

Table 7.4-1. Inspection Practices for Marine Infrastructure

Infrastructure Type	Analogous Industry	Practices and Guidelines
Water Front Facilities	Petroleum and Shipping	*Marine Safety Manual Volume II: Materiel Inspection, COMDTINST 16000.7B,* US Coast Guard, 2014 *Waterfront Facility Compliance Booklet,* Department of Homeland Security, U.S. Coast Guard, CG-5562A (Rev 6-04) *Section 3102F – Audit and Inspection,* California State Lands Commission (CSLC) Code 31F – Marine Oil Terminals, Div. 2.

The use of checklists during maintenance and inspection activities is a good practice to ensure that inspections are thorough. Table 7.4-2 is a sample checklist for marine infrastructure.

Table 7.4-2. Example Checklist for Marine Infrastructure

Marine Facility- Above Water Equipment	Y	N	N A	Remarks
1. Are there defects in loading arms that allow discharge of oil or hazmat?	☐	☐	☐	
2. Are there defects in loading piping that allow discharge of oil or hazmat?	☐	☐	☐	
3. Are there unrepaired kinks, bulges, soft spots or other defects in transfer hoses?	☐	☐	☐	
4. Are there cuts, slashes, or gouges that penetrate the first layer of hose reinforcement?	☐	☐	☐	
5. Are Remotely Operated Valves (ROVs), tank level alarms, and ESD systems operating properly?	☐	☐	☐	
6. Are Vapor Control System (VCS) and flame arrestors inspected internally annually?	☐	☐	☐	
Marine Facility- Above Water Structures	**Y**	**N**	**N A**	**Remarks**
7. Is there evidence of collision damage on mooring or breasting dolphins?	☐	☐	☐	
8. Is there concrete spalling of dolphins or jetty structure?	☐	☐	☐	
9. Is rebar visible on deteriorated reinforced concrete structures?	☐	☐	☐	

Table 7.4-2. Example Checklist for Marine Infrastructure, continued

	Y	N	N A	Remarks
10. Is there advanced deterioration or overstressing observed on widespread portions of the structure, and has an engineering evaluation verified that the deterioration does not significantly reduce the load bearing capacity of the structure? (i.e., the capacity of the structure is no more than 25 percent below the structural requirements of applicable standard).	☐	☐	☐	
11. Is there advanced deterioration, overstressing or breakage that may have significantly affected the load bearing capacity of primary structural components, based an engineering evaluation that determined the capacity of the structure is more than 25 percent below the structural requirements of applicable standard. (Not fit for purpose, local failures are possible and loading restrictions may be necessary).	☐	☐	☐	
Marine Facility - Underwater Structure	**Y**	**N**	**N A**	**Remarks**
12. On steel structures is there moderate mechanical damage including corrosion pitting and loss of section?	☐	☐	☐	
13. On concrete structures is there major spalling and cracking, rust staining, exposed reinforcing steel and/or pre-stressing strands?	☐	☐	☐	
14. On timber structures is there major loss of section, broken piles and bracings, severe abrasion or marine borer attack?	☐	☐	☐	
15. Is there coating integrity and effectiveness failure on coated steel components?	☐	☐	☐	
16. Is there evidence of significant damage or loss of effectiveness of wraps on steel, concrete or timber wrapped components?	☐	☐	☐	
17. Is there evidence of significant damage or deterioration of the underlying component on steel, concrete or timber encased components?	☐	☐	☐	
18. Is there joint sealant damage at transverse joints (including extrusion, hardening, adhesive failure (loss of bond), cohesive failure (splitting), or complete loss of the sealant. The presence of weed growth in the joint is also an indication of joint seal damage)?	☐	☐	☐	
19. Has the ship size limit been increased since initial installation?	☐	☐	☐	
20. Are there large accumulations of underwater vegetation that inhibit inspection of structural components and promote Corrosion Under Deposits (CUD)?	☐	☐	☐	
21. Has the fire seawater intake line corroded at the tidal interface?	☐	☐	☐	

7.4.3   Marine Facilities Aging Warning Signs

Often there are warning signs that precede a potential catastrophic failure. Some examples for marine facility infrastructure are listed below.

- Exposed reinforcing steel on concrete structures
- Damaged structural components due to ship collision or hard berthing events
- Visual and documented lack of maintenance of top side equipment including:
- fire suppression system
- spill containment, sumps and sump tanks
- valve and manifold labeling
- loading arm swivel joints
- sub deck piping (out of sight, out of mind)
- transfer hoses
- lifting crane
- Overdue inspection of underwater structures

The following case study reflects the aging concerns of a supertanker named Betelgeuse.

Betelgeuse was a first-generation oil supertanker built in 1968 used to ship Middle East crude to refineries in Western Europe. In late 1978, and due to severe weather, the ship was diverted to the Whiddy deep water terminal in Western Ireland. En-route to the terminal the ship sustained some minor damage causing oil to leak from the bulkheads, a possible warning sign of future problems. The origin of the leaks was determined and temporary repairs were made.

The terminal had been constructed a decade earlier but due to neglect its firewater systems had deteriorated and chronic leakage problems became common. To reduce operating costs, the firewater systems were placed in standby mode only to be commissioned when needed.

Oil transfer operations commenced on the morning of January 8, 1979. A sequential unloading procedure is required to ensure that oil is evenly removed from each of the storage compartments thereby reducing bending stresses on the hull of the aging ship. With little warning the hull began to crack and loud noises were heard followed by fires on the upper deck. The crew abandoned the ship and mustered on the jetty. With no firewater available, the crew anxiously awaited rescue from an emergency vessel. Before the emergency rescue boat arrived a series of violent explosions erupted causing the hull of the ship to break in half. Fifty people were killed including crew members and rescue workers.

This event serves to illustrate the importance of maintaining all critical facilities and infrastructure in reliable condition and a state of readiness. For more information related to the Betelgeuse case study, please refer to Appendix A.

7.5   UNDERGROUND UTILITY SYSTEMS

This section addresses standards and practices specific to Underground Utility Systems that support the aging infrastructure principals presented hither to.

7.5.1 Electric Cables

Electric Cable Installation Information. The following is a list of some typical information pertaining to private underground electric cable systems that should be compiled. It is intended to cover plant site and facility underground power cables that are beyond the utility company's substation and jurisdiction.

Codes and Standards. These standards are primarily derived from the power industry including selecting, installation, maintenance, and repair practices, and are generally applicable to private systems. Some examples that may apply to plant underground infrastructure are presented below:

- IEEE 1242 - *Guide for Specifying and Selecting Power, Control, and Special Purpose Cable for Petroleum and Chemical Plants*, Institute of Electrical and Electronic Engineers, 1999. This Guide addresses wire and cable design, materials, testing and installation
- IEEE P1185 - Recommended Practice for Cable Installation in Generating Stations and Industrial Facilities, Institute of Electrical and Electronic Engineers, 2010
- API RP 540 Electrical Installations in Petroleum Processing Plants, American Petroleum Institute

Design Information

- Cable size and kVA rating
- Manufacturer and fabrication materials
- Type of burial (direct, conduit, other)
- Depth of burial
- Termination details
- Other utility crossings
- Subsurface junction or vault boxes
- Access points
- Fault protection

Drawings

- Underground utility plot plan showing power cables and other services
- Trench cross section detail
- Subsurface junction or vault box details
- Termination wiring diagram

Other

- Installation field inspection records
- Repair history
- Accident history (external intrusion)

Electric Cable Inspection. Inspection of underground cable systems continues to be an area of focus. Some examples of inspection practices for underground cable systems are listed in Table 7.5-1. As explained in the second reference, the accuracy of available diagnostic techniques is better at detecting good conditions than bad.

Table 7.5-1. Inspection Practices for Underground Cable Systems

Infrastructure Type	Analogous Industry	Practices and Guidelines
Power Cable	Power Industry	400-2012 - IEEE *Guide for Field Testing and Evaluation of the Insulation of Shielded Power Cable Systems Rated 5 kV and Above, IEEE, 2012* *Diagnostic Testing of Underground Cable Systems (Cable Diagnostic Focus Initiative),* National Electric and Energy Testing and Research Center (NEETRAC), Georgia Tech Research Corp., DOE Award No. DE-FC02-04CH11237, 2010. A comprehensive report on failure modes and diagnostic techniques for testing underground power cables.

Most available checklists for electrical systems rely on visual inspection of above ground installed electrical components or inspection activities during construction. The latter is very important as most faults in underground cabling are a result of damage during construction. For deterioration, due to aging, using one of the diagnostic techniques described in Table 7.5-1 should be considered. Table 7.5-2 is a sample checklist for underground electric cables.

Additionally, an insulation resistance tester, also referred to as a megger can be used to determine the condition of cable insulation for underground installations. This tester is a portable instrument that provides readings of insulation resistance, to detect both good and bad insulation. It is recommended to conduct periodic testing as part of the preventive maintenance, and analyze the measurement trends. (Megger, 2006)

Table 7.5-2. Example Checklist for Maintenance and Inspection of Underground Cable Systems

Underground Electrical Cables	Y	N	NA	Remarks
1. Are there local power outages after it rains?	☐	☐	☐	
2. Are ground fault protectors tripping frequently?	☐	☐	☐	
3. Is there evidence or history of excessive conductor current?	☐	☐	☐	
4. Have petrochemical spills, transformer oil leaks or fertilizer spills been reported above underground cable routes (causing swelling, melting or cracking of insulation and outer jacket sheathing)?	☐	☐	☐	
5. Is unjacketed cable buried in soil that enhances copper corrosion (causing loss of neutral)?	☐	☐	☐	
6. Are power cable routes known and extensively mapped with effective administrative controls to reduce the risk of external intrusion (digging up) of power cables?	☐	☐	☐	

7. Is there deteriorated ......?	☐	☐	☐	

Underground Cable Aging Warning Signs. Often there are warning signs that precede a potential catastrophic failure. Some examples for underground cable aging warning signs are listed below.

- Degradation of cables due to moisture, high humidity or contaminants
- Degradation of cables caused by vibration
- Loss of dielectric isolation sufficient to cause functional failure of the circuit it is used in

7.5.2   Utility Underground Piping: Fuel Gas, Cooling Water, Fire Water, Drains and Sewers

Utility Underground Piping Information. The following is a list of some typical information pertaining to underground utility piping systems that may be applicable to plant systems.

Codes and Standards

- ASCE 15-98 Standard Practice for Direct Design of Buried Precast Concrete Pipe Using Standard Installations (SIDD), American Society of Civil Engineers, 2000
- Design Data #40—Standard Installations and Bedding Factors for the Indirect Design Method, American Concrete Pipe Association (ACPA), 1996
- *Concrete Pipe Design Manual*, American Concrete Pipe Association (ACPA), 2000
- M086MM086-15-UL, *Standard Specification for Non-Reinforced Concrete Sewer, Storm Drain, and Culvert Pipe*, American Association of State Highway and Transportation Officials (AASHTO), 2015
- ASME/ANSI B31.8 - *Gas Transmission and Distribution Piping System*, Code of Pressure Piping, American National Standards Institute (ANSI)
- 49 CFR Part 192, Subpart G - General Construction Requirements for Transmission Lines and Mains, DOT PHMSA
- General Specifications for Gas Service, State and Local Authorities with Jurisdiction
- *Guidelines for the Design of Buried Steel Pipe*, American Lifeline Alliance, (a public-private partnership between FMEA and ASCE), 2001
- NFPA 24: Standard for the Installation of Private Fire Service Mains and Their Appurtenances, National Fire Protection Association, 2016 Edition
- NACE SP0169-2013 Standard Practice, *Control of External Corrosion on Underground or Submerged Metallic Piping Systems*, National Association of Corrosion Engineers (NACE), 2013

Design Information

- Pipe process design parameters and mechanical design specifications

- Pipe size
- Burial Depth and bedding factors
- Gradient for gravity drains
- Pipe system stress analysis
- Pipeline construction bedding and overburden material
- Pipe coating and cathodic protection system
- Joint sealing system
- Sewer/drain system design information

Drawings

- Layout and mapping of underground piping system
- Trench cross section detail
- Valve box and interceptor details

Other

- Soil and geotechnical information
- Seismic design conditions (if applicable)
- Installation field inspection records

Utility Underground Piping Inspection. Several professional societies and associations have maintenance and inspection practices for utility underground piping infrastructure. Examples of inspection practices for utility piping infrastructure are listed in Table 7.5-3.

Table 7.5-3. Inspection Practices for Underground Utility Piping Infrastructure

Infrastructure Type	Analogous Industry	Practices and Guidelines
Steel Piping	Gas Transmission	API RP 1110 *Recommended Practice for the Pressure Testing of Steel Pipelines for the Transportation of Gas, Petroleum Gas, Hazardous Liquids, Highly Volatile Liquids, or Carbon Dioxide,* American Petroleum Institute, 5th Edition, 2015
		NACE SP0102-2010 Standard Practice, *In-Line Inspection of Pipelines,* National Association of Corrosion Engineers (NACE), 2010
		NACE SP0169-2013 Standard Practice, Control of External Corrosion on Underground or Submerged Metallic Piping Systems, Section 10 O&M, National Association of Corrosion Engineers (NACE), 2013
Non-Metallic Piping	Municipal Water and Sewer	*Nondestructive, Noninvasive Assessment of Underground Pipelines,* Michael Dingus et. al, American Water Works Association (AWWA) Research Foundation
		Pipe Condition Assessment Using Cctv: Performance Specification Guideline, National Association of Sewer Service Companies (NASSCO), 2014.

Table 7.5-3. Inspection Practices for Underground Utility Piping Infrastructure, continued

Fire Mains	Fire Protection	NFPA 24: *Standard for the Installation of Private Fire Service Mains and Their Appurtenances*, National Fire Protection Association, Chapter 14: Inspection, Testing & Maintenance, 2016
		NFPA 291: *Recommended Practice for Fire Flow Testing and Marking of Hydrants*, National Fire Protection Association, 2016
		FM Global Loss Prevention Data Sheet 3-10
		Private Service Water Mains and Data Sheet 2-81
		Fire Protection System Inspection, Testing and Maintenance and Other Fire Loss Prevention Inspections

The use of checklists during maintenance and inspection activities is a good practice to ensure that inspections are thorough. Defects in underground utility infrastructure are generally not directly visible. Table 7.5-4 is a sample checklist for some indirect evidence of aging of underground utility piping infrastructure.

Table 7.5-4. Example Checklist for Underground Utility Piping Infrastructure

Fuel Gas Piping	Y	N	N A	Remarks
1. Are there signs of surface frosting along pipeline route?	☐	☐	☐	
2. Is there detectible odor alone the pipeline route? Note: Portable sampling instruments are available for leak testing.	☐	☐	☐	
3. Are pipeline Right of Way (ROW) markers provided for Outside Battery Limits (OSBL) buried sections?	☐	☐	☐	
4. Is there Cathodic Protection (CP) for the pipeline?	☐	☐	☐	
Fuel Gas Piping	**Y**	**N**	**N A**	**Remarks**
5. Was there a CP survey done within the last year?	☐	☐	☐	
6. Were any interferences at other pipeline crossings found during the CP survey? Actions taken?	☐	☐	☐	
7. Internal pigging has been performed to evaluate pipeline integrity / corrosion with results trended (where possible)?	☐	☐	☐	
8. Are water crossings are identified	☐	☐	☐	
9. Are Aerial crossing identified and the integrity of aerial crossing structures adequate?	☐	☐	☐	
Sewers and Drains	**Y**	**N**	**N A**	**Remarks**
10. Are storm drains backing up during heavy rain?	☐	☐	☐	

| 11. Are sewers and drain lines protected from heavy vehicle loads? | ☐ | ☐ | ☐ | |

Table 7.5-4. Example Checklist for Underground Utility Piping Infrastructure, continued

Sewers and Drains	Y	N	N A	Remarks
12. Are there ground water monitoring wells alone closed drain system underground lines?	☐	☐	☐	
13. Is there evidence of groundwater contamination from monitoring well data?	☐	☐	☐	
Fire Water Mains	**Y**	**N**	**N A**	**Remarks**
14. Was a flow test performed in accordance with NFPA 291?	☐	☐	☐	
15. Have any flow test follow-up actions been complete?	☐	☐	☐	
16. Are hydrants classified in accordance with their rated capacities?	☐	☐	☐	
17. Is the hydrant class indicated at each location?	☐	☐	☐	
18. Is the fire water system periodically flushed?	☐	☐	☐	
19. Occurrence of surface boils or damp stops during flow test?	☐	☐	☐	
20. Are all fire system valves in satisfactory condition and locked in the open position to prevent inadvertent isolation of the fire water system?	☐	☐	☐	
21. Are the fire system valves inspected and exercised on a regular basis (weekly/monthly/annually) to verify operational integrity?	☐	☐	☐	
22. Are the fire pumps in automatic operating mode, in satisfactory operating condition and tested weekly and annually?	☐	☐	☐	

A die test is one of the common ways to determine location of a suspected leak for cooling water and fire water lines as well as sewer and drain systems.

Underground Utility Piping Aging Warning Signs

- Declining fire water capacity and pressure
- Increasing levels of chemical contaminates in well data
- Noticeable odor along fuel gas piping route
- Visible liquid pools
- Visible concrete degradation (if applicable)

8

DECOMMISSIONING, DISMANTLEMENT AND REMOVAL OF REDUNDANT EQUIPMENT

8.1   INTRODUCTION

It is important to understand the difference between aging and derelict facilities. Facilities which have long been abandoned or which have been neglected for a considerable period may be beyond repair and should not be considered for future service. Attempting to restore such facilities may be impractical and uneconomical and could introduce new safety concerns. Such facilities should be demolished and potentially contaminated debris sent to an approved waste site.

This chapter addresses hazards posed when attempting to demolish and dispose of facility assets that have exceeded their economic and physical life expectancy. Infrastructure and facilities that have reached the end of their lifecycle or that are deemed unsafe to operate should be permanently retired from service. However, retirement is only one of several steps, which should normally include de-commissioning, sterilization (cleaning), dismantlement, demolition and disposal.

Figure 8.1-1. Image of Building Awaiting Demolition

8.2   EQUIPMENT HAZARDS

8.2.1   Unknown or Undocumented Condition

Often equipment that has been retired from service is left in place. Facilities left in an inactive state without proper post-retirement decommissioning can involve risks, since they may still pose a hazard and a liability. Idle assets are a breeding ground for biohazards and occasional trespassers who might sustain injuries.

Once the asset has been retired, systematic inspection and record keeping generally ceases. If the necessary steps to passivate chemical hazards and ensure structural integrity going forward are not taken at the time of retirement, the condition on the asset may be unknown when dismantling and disposal are finally undertaken. This may lead to exposure to hazardous material by personnel involved in dismantling, or to the public from improperly classified hazardous waste. Additionally, the structural integrity of the asset may seriously deteriorate by the time a decision to dismantle is finally made, posing an increased risk of structural failure and injury to demolition workers.

8.2.2   Dismantling Residual Chemical Hazards

Old infrastructure, particularly process buildings, can contain hazardous construction materials no longer allowed or contaminated, such as:

- Asbestos pipe insulation
- Refractory insulating materials
- Transite (asbestos – cement) pipe, ducts and hoods
- Transite shingles and siding
- PCB containing electrical components
- Caulking and glazing materials
- Sanitary and chemical drain pipe
- Wood, cement and tile floors contaminated with production materials
- Silicate containing materials

If not identified prior to dismantling and properly managed, materials such as these can pose a serious health risk to workers involved in the dismantling activities. It could also result in fines, since there are common safety regulations that must be followed for handling some of these materials. Also, additional schedule and cost impacts may incur if found during a dismantling project.

When little information is available about the asset's last in-service conditions and material processed, sampling and analysis must be employed to obtain important information needed to plan and execute a safe decommissioning plan as seen in Section 8.3.1.

Lessons From the Nuclear Industry. Public acceptance of nuclear power plants has diminished because of several uncontrolled natural disasters in Europe and Asia in recent years. Consequently, fewer nuclear plants are now being planned or built in the developed nations. Some European countries are phasing out nuclear power altogether. The de-commissioning and final disposition of existing nuclear power plants presents an economic and environmental challenge.

There are currently over four hundred nuclear reactors in service mainly in North America, Europe and Asia. These are typically configured in groups of two to four within power plants. Many of these were built and commissioned in the latter part of the last century. Even with latest technology and quality construction, these plants require periodic outages for refueling and boiler tube replacement. This costly maintenance is required to ensure that the plants perform reliably. With an anticipated life expectancy of 40 years, many of these plants are now reaching the end of their useful lifecycle. In fact, regulatory authorities that oversee the licensing of nuclear power installations often stipulate full retirement after four decades in operation.

The retirement of a nuclear facility is both complex and costly. Used fuel must be disposed of in a safe manner and quarantined in a burial site remote from populated areas. All equipment and supporting infrastructure must be de-commissioned, de-contaminated and dismantled with care taken to separate out components that have been exposed to measurable levels of radiation. The disposal of each component must be managed, that is, a record must be kept of how it is handled and where it is disposed of. Ultimately, the risk of accidental or unauthorized contact with a radioactive source must be near zero. The infrastructure itself including buildings and foundations is usually covered over and ground access permanently restricted. While these measures may vary with location and jurisdiction, they serve to illustrate the degree of diligence required when long-term public safety is a concern.

This hazard also exists with decommissioning offshore platforms where nucleonic level gauges have been left to decay naturally in the ocean. This is an issue for North Sea platforms because ALARP cases were presented and accepted by the UK HSE

The challenges of dealing with aging nuclear facilities are obviously unique. Long-term health and safety hazards can remain for decades (or sometimes centuries) owing to the half-life of radioisotopes. For this reason, the nuclear industry cannot afford to take chances. Risk is a never-ending phenomenon that must be managed on a daily basis, there are no time-outs or intermissions.

Few readers are likely to be directly involved in the nuclear industry. Nonetheless, there may be some lessons from the nuclear sector that can be shared with industry at large. These can be presented in the form of questions.

- As the end of the facility lifecycle approaches, are proper repairs made with quality replacement parts?
- Are short cuts ever taken on the premise that an accident is unlikely to occur in the brief time period remaining?
- Can we be certain that used and worn components do not fall into the wrong hands?
- How can we take responsibility for used materials that are disposed of through scrap dealers?
- How can we be sure that used materials are not misrepresented and put into secondary service?
- Are derelict sites periodically inspected to ensure that hazards are not openly exposed to the atmosphere or to wildlife (wind or erosion)?
- Are aquifers tested to ensure that hazardous materials have not leached into the subsoil below?

8.2.3   Custody After Removal

The plant's responsibility does not end once the demolition debris leaves the site. Responsible care is required to ensure there are no residual liabilities resulting from inappropriate actions due to third party handling of debris and waste. Care should be taken to ensure that contractors:

- Dispose of retired equipment to a designated waste site or burial ground to avoid soil and groundwater contamination
- Take precautions during transport of debris to avoid exposing the public to hazardous materials along roads and highways
- Secure used or scrapped equipment to prevent retrieval and reuse by illegal operators'

To safeguard against future liabilities by unscrupulous vendors, careful due diligence is necessary in the selection of third party contractors including:

- Reviewing qualifications and relevant past projects (client list)
- Contacting references
- Reviewing testimonials
- Searching history for regulatory violations

Remember the adage "You get what you pay for and be wary of low-ball prices. It is important to do it right, even if the direct business benefit is hard to quantify.

8.3   FINAL DECOMMISSIONING PRACTICES

Facilities ultimately should be decommissioned in a safe and responsible manner using documented procedures to isolate, remove and dispose of obsolete decommissioned, and/or unneeded equipment so they pose no further risk. Changes need to be managed using the MOC program, with any process related hazards evaluated. Asset integrity program records and plant P&ID's will need to be revised as needed so that documentation aligns with the plant conditions so that they pose no further risk. This section describes some of the additional considerations required to achieve this objective.

8.3.1   Cleaning

The purpose of cleaning is to remove, neutralize and/or otherwise passivate any process material that could pose a future health risk, or could cause continued aging that might lead to partial or total failure of the asset. Soon after the asset has been permanently removed from service, it should be scheduled for cleaning while knowledge and documentation of its last service are known. The specific methods for cleaning will depend on what needs to be removed. For infrastructure, it can be as simple as flushing drains with water, or a solution of detergent, or mild caustic. For a building containing asbestos material, it could require engaging a contractor to identify, label and encase asbestos containing components as a minimum. Prior to dismantling, these materials need to be removed and properly disposed. A plan for the cleaning process needs to be developed taking account of:

- Type and hazards of the contaminating materials and any cleaning fluids
- Health risks and required personnel protective equipment
- Work permits required
- Cleaning methods and means to verify effectiveness
- Cleaning materials and equipment required
- Accessing the infrastructure including digging or lifting equipment
- Expected duration of cleaning effort
- Disposal of cleaning residues
- Use and management of third party contractors
- Estimated cost

Since this will be a non-routine activity, performing a Job Safety Analysis (JSA) should also be included as part of the plan.

The cleaning process should be scheduled and documented through the plant's maintenance management, work order, and record keeping process. However, depending on the size and complexity of the activity, handling the activity as a special project may be necessary for proper management.

8.3.2   Retaining Spare Equipment and Parts

It may be desirable to retain some equipment components in decommissioned facilities. For example, electrical components which are only available from the used equipment market, but are still in use in other areas of the plant. These items should be removed soon after decommissioning to allow cleaning, refurbishment and storage to minimize additional aging. Items should be cataloged in the spare parts inventory systems including descriptors (age, size, model #, rating, capacity, etc.), designated as use/refurbished with a brief history of prior service, and time in use.

Before being accepted for retaining as spare equipment/part, the item needs to be thoroughly inspected and tested to verify its functionality and worthiness for continued use. Depending on the type of asset, the testing may include a pressure test, a mechanical or electrical function test, or some ASTM test procedure such as a dielectric strength for power cable insulation, or flaws in precast concrete pipe.

8.3.3   Disposal of Chemicals

During the decommissioning process, several sources of chemical residues may be present. One is any residual production materials remaining in equipment after de-inventorying prior to removal from service. Another is cleaning materials and fluids that are contaminated with production materials. There can also be construction materials such as asbestos that require proper disposal. The cleaning and decommissioning plan needs to identify and classify materials as hazardous or non- hazardous, and develop specific disposal requirements for all hazardous wastes. The proposed disposal procedures need to be in compliance with applicable environmental and public health regulations.

While the presence of Naturally Occurring Radioactive Material (NORM) is not typical (albeit, manufacturing cement blocks with radioactive sand has occurred) for infrastructure assets, NORM and Technologically Enhanced Naturally Occurring Radioactive Material (TENORM) may also be present in electrical equipment or in materials concentrated in pipes, storage tanks or other filtration equipment used in water treatment. The contamination may also be present in mineral scale, sludge, or evaporation ponds or pits (Clean Harbors, 2016).

NORM was not subject to regulatory control under the Atomic Energy Act of 1954 or the Low Level Radioactive Waste Policy Act. Thus, NORM was subject primarily to individual state radiation-control regulations. Section 651(e) of the Energy Policy Act of 2005 gives the NRC (US National Research Council) jurisdiction over discrete sources of NORM by redefining the definition of source material, (TCEQ, 2015).

The NRC exercises regulatory authority over radioactive materials in regions or jurisdictions that do not have specific agreements. The NRC and the specific jurisdictions coordinate the regulation of radioactive materials through the National Materials Program. Toward that end, the NRC retains a leadership and oversight role in the program through the Integrated Materials Performance Evaluation Process (IMPEP). In particular, IMPEP ensures uniform nationwide regulation by reviewing the regulatory performance of both the NRC and the States using a common set of performance criteria (NRC, 2015). Therefore, when NORM is encountered, it will require compliance with specific state regulations if the plant is located in one of the 37 Agreement States. The best practice is to contact a licensed hazardous waste disposal firm that also handles NORM.

Pyrophoric materials (combusts on contact with air) also require special handling. It is important to make sure such materials are rendered inert, before attempting to transport them on public roads and highways. This can generally be accomplished by exposing the material to air for a sufficient length of time, at a safe location, away from process and flammable storage areas.

8.4   DISMANTLING AND DISPOSAL

8.4.1   Degassing

Dismantling natural gas and fuel gas pipelines will require freeing the lines of flammable gas and inerting before disassembly. After depressurization, the remaining gas in the pipeline needs to displaced with an inert (oxygen free) fluid. The typical choices are low pressure steam or nitrogen. When steam is used, the line should be dried out using compressed air, after all traces of flammable hydrocarbons are gone. Nitrogen purging of the gas, when available, is more efficient as it can be left in the line for inerting and moisture control.

8.4.2   Inerting

Blanketing of metal equipment and piping with an oxygen and moisture free gas is usually performed to protect it from internal corrosion, when it is left out of service for extended periods of time. Nitrogen is the gas of choice, due to its very low dew point and oxygen concentration. The types of infrastructure equipment where inerting would be appropriate include, natural and fuel gas piping and knock out drums, instrument and plant air receivers and piping. Inerting may not be necessary if dismantling of the equipment is scheduled to occur shortly following degassing.

Both initial blanketing and re-entry to the inerted equipment has to be done by following a very careful procedure to do the job safely. Asphyxiation is an extremely dangerous consequence of not following appropriate safety protocols. One of the leading causes of industrial fatalities is over exposure to inert gases.

8.4.3   Removal from Operating Facilities

Removal of decommissioned equipment usually requires the use of some type of rigging, hoisting, digging, and hauling equipment. The removal process becomes potentially more hazardous when the removal site is inside or in close proximity to in-service equipment. Of particular concern is the lifting of heavy objects near or over live operating equipment. There have been several significant chemical plant incidents due to dropped objects or toppled crane booms causing damage to process equipment and piping, or injuries to demolition personnel.

Over 50% of all mobile crane accidents are the result of mistakes made while the crane was being set-up (IHSA, 2012). Most all of these accidents could have been prevented by following the manufacturer's recommendations for assembly and dismantling, by using the correct components, and by observing safety precautions. The cited reference provides a pre-job checklist for avoiding mishaps while operating lifting equipment in hazard areas (IHSA, 2012). While this checklist was designed for crane operation, some of the precautions and safety considerations are generally applicable to any large construction equipment operating near process or utility hazards.

8.4.4   Site Cleanup

While infrastructure facilities and systems are ancillary to the main production equipment, they can still contain materials that are considered health risks. For example, certain waste water residues from decommissioned retention ponds and water treatment facilities may be considered covered under Resource Conservation and Recovery Act (RCRA) regulations, which is the main federal regulation that may apply to the site cleanup of solid waste. This regulation defines solid wastes both non-hazardous and hazardous. The EPA in conjunction with state Departments of Environmental Protection implements this law. In some cases, the state regulations may exceed the federal requirements.

The Comprehensive Environmental Response, Compensation, and Liability Act of 1980 (CERCLA) is another United States federal law designed to clean up sites (designated as Superfund sites) contaminated with hazardous substances and pollutants. CERCLA was enacted by Congress in 1980 in response to the threat of hazardous waste sites, typified by the Love Canal disaster in New York. It authorizes federal natural resource agencies, primarily the Environmental Protection Agency, states and Native American tribes to recover natural resource damages caused by hazardous substances, though most states have and most often use their own versions of CERCLA. Other jurisdictions worldwide are likely to enact similar legislation in the future.

The first step in site cleanup is to determine whether there are any known or suspected materials present that may be covered by federal and state environmental regulations. The regulations mentioned above are not the only ones that may apply, and it is not the intent of this treatise to provide a complete summary of the potentially applicable regulatory setting. Suffice it to say that once the nature of the contamination is determined, it will be necessary to review the regulatory environment to establish what statues apply. Surface and groundwater contamination are also major issues that may need to be addressed.

When site contamination has been encountered and characterized, the next step is to select methods to accomplish cleanup. Removing pollution or contaminants from groundwater, surface water, or soil involves environmental remediation. To get the job done successfully, it is necessary to first understand the different cleanup methods and how they work (Rodewald, 2014). Here is a list of some applicable methods:

- Ground Water Pumping and Treatment: This method involves extracting ground water with a vacuum pump, and then separating contaminants with techniques like carbon adsorption, biological treatment, and air stripping
- Waste Water Treatment: A method for removal of contaminates from waste water with techniques such as physical separation, chemical treatment, and biological treatment
- Bio-remediation: Employing natural bio-degradation of contaminants by micro-organisms, which can be enhanced through the addition of nutrients or cultivation
- Incineration: The use of extremely high temperature to destroy organic compounds contained within hazardous waste
- Thermal Desorption: This method utilizes high temperatures to heat contaminated soil, vaporizing volatile and semi-volatile organics (like mercury or hydrocarbon), which are then either collected or treated with an afterburner
- Removal and Disposal: This method involves the physical removal of contaminated equipment, soil, water, sludge and/or tanks and transporting it to an approved hazardous waste disposal facility

In many cases for infrastructure sites, the last option may be preferred, if and when contamination is found. Employing an experienced environmental remediation contractor to manage the cleanup can be a consideration when in-house environmental expertise and resources are limited. To carry out a proper site cleanup requires planning and potential coordination with state agencies. This may require third-party resources.

In the case of plant closure and demolition of all facilities at a production site, provisions should be made to ensure that future land developments do not encroach on the abandoned plant area or expose industrial hazards that have not been t remediated completely.

Important reasons for addressing and proceeding with site cleanup soon after demolition is completed include:

- Known liabilities can affect the future value of an asset when a company decided to sell the business or the property after termination of operations

- If hazardous materials are leaching into the soil, the situation can worsen the longer the issue is not addressed

8.4.5 Scrap Value

Some materials retain a significant scrap or reclaim value even as they reach the end of their useful service life. Fired heater tubes in high temperature cracking service are one such an example. These are sometimes made with HK40 or HP45, both high temperature resistant cast alloys. The reclaim value of such tubes can represent a significant portion of the original cost. This issue introduces a further complexity into the risk question "should we retain the facilities and/or equipment in a backup contingency mode or should we recover the scrap value now?". The optimum strategy will depend on several factors such as the likelihood of a need for backup contingency, the market criticality of the operation, and how monetary recovery might be taxed.

9

ONWARD AND BEYOND

It is hoped that you found this book to be interesting and that it will prove to be useful in dealing with the challenges of aging facilities and infrastructure. However, even the best concepts and ideas do not translate into results unless someone initiates a course of action. You, as a reader, have an opportunity to communicate what you have learned to others in your organization and make them aware of the vulnerabilities that might reside at their facilities. Codes and regulations do not necessarily address high risks outside principal operating areas. Pipe racks, roadways, loading racks and warehouses are often neglected since they are not seen as critical to the business. Nonetheless, a surprise failure can totally cripple an operation or lead to catastrophe. Pressure to maximize production and minimize operational costs sometimes drives maintenance deferments that contribute to premature aging. In addition, infrastructure priorities are often in the shadow of production assets and may be ignored or overlooked.

What can you do and how can you do it? The following strategy is one of many that can get things moving.

1. Flag the issue to your colleagues and to area management. Don't raise a concern unless you know for sure that you have facilities that may fit into one or more of the categories listed in this concept book. A short presentation to supervision and management highlighting some of the concerns herein should capture their attention. Use case studies and photos if applicable to your type of operation.
2. Suggest that someone check equipment data archives to determine when facilities and infrastructure were initially built and when they were last inspected. Were any significant concerns or trends noted and how were these addressed?
3. Develop some convincing arguments for Inspection and Maintenance functions with the aim of at least looking at facilities outside the normal bounds of maintenance planning. This initiative should start off small. It is unlikely that you will mobilize a large inspection crew in the early stages and you could lose credibility. However, if major problems are uncovered, the inspection initiative should gain momentum.
4. Extend formal risk assessment protocols outside principal operating areas. Most companies with mature process safety programs conduct a formal guideword HAZOP or FMEA (Failure Modes and Effects) on hazardous operating facilities at 5 year intervals. Consider extending the scope of these studies outside traditional boundaries using a "What If" analysis.
5. Assign priorities to significant concerns raised in items 2 to 4 above. Maintenance budgets will likely need to be adjusted to address this added work scope.

Within the principal production areas of an operation there are some valuable opportunities arising from the use of this book. Both HAZOP and FMEA use traditional guidewords and equipment failure modes. The concept of aging as it relates to continuous exposure to harsh conditions can suggest new guidewords and failure modes. This book can better prepare a team to participate in a facility risk assessment.

ACRONYMS

AASHTO	American Association of State Highway Transportation Officials
ACI	American Cement Institute
ACPA	American Concrete Pipe Association
AD	Airworthiness Directives
AHA	Asset Hazard Analysis
AIChE	American Institute of Chemical Engineers
AIM	Asset Integrity Management
AISC	American Institute of Steel Construction
ALARP	As Low as Reasonably Practicable
ALCM	Asset Lifecycle Management
AMCA	Air Movement and Control Association
ANSI	American National Standards Institute
API	American Petroleum Institute
ASCE	American Society of Civil Engineers
ASHRAE	American Society of Heating, Refrigerating, and Air-Conditioning Engineers
ASME	American Society of Mechanical Engineers
ASNT	American Society of Non-Destructive Testing
ASTM	American Society for Testing and Materials International
AWWA	American Water Works Association
BCTC	British Columbia Transmission Corporation
C&I	Control and Instrumentation
CAPEX	Capital Expenditure
CCPS	Center for Chemical Process Safety
CERCLA	Comprehensive Environmental Response, Compensation, and Liability Act
CM	Corrective Maintenance
CMMS	Computerized Maintenance Management System
CNC	Computer Numeric Control
CP	Cathodic Protection
CPI	Chemical Process Industries
CSB	Chemical Safety Board
CSLC	California State Lands Commission
CUD	Corrosion Under Deposits
CUF	Corrosion Under Fireproofing
CUI	Corrosion Under Insulation
DOT	Department of Transportation

DTBT	Ductile Transition to Brittle Temperature
EC&I	Electrical, Control and Instrumentation
EHS	Environmental Health and Safety
EIT	Engineer-in-Training
EPA	Environmental Protection Agency
ER	Emergency Repair
ESD	Emergency Shutdown
EUAC	Expected Uniform Annual Cost
FAC	Flow-Assisted Corrosion
FCC	Fluid Catalytic Cracking
FEMA	Federal Emergency Management Agency
FFS	Fitness for Service
FHWA	Federal Highway Administration
FIT	Failure in Time
FMECA	Failure Modes Effects and Consequence Analysis
FMEDA	Failure Modes and Effects Diagnostic Analysis
FTA	Fault Tree Analysis
HP	Horse Power
HSE	Health and Safety Executive
HTHA	High-Temperature Hydrogen Attack
HVAC	Heating, Ventilation and Air Conditioning
I&M	Inspection and Maintenance
IBC	International Building Code
ICC	International Code Council
IEEE	Institute of Electrical and Electronics Engineers
IMC	International Mechanical Code
IMPEP	Integrated Materials Performance Evaluation Process
IPL	Independent Protection Layers
ISA	Instrument Society of America
JSA	Job Safety Analysis
KPI	Key Performance Indicators
LOPA	Layer of Protection Analysis
LPG	Liquefied Petroleum Gas
MARS	Major Accident Reporting System
MAWP	Maximum Allowable Working Pressure
MAWT	Maximum Allowable Working Temperature
MCC	Motor Control Center
MMS	Maintenance Management System
MOC	Management of Change
MOP	Manuals of Practice
MOTEMS	Marine Oil Terminal Engineering & Maintenance Standards

MTTF	Minimum Time to Failure
NACE	National Association of Corrosion Engineers
NASSCO	National Association of Sewer Service Companies
NDE	Non-Destructive Examination
NEC	National Electric Code
NECA	National Electrical Contractors Association
NEETRAC	National Electric and Energy Testing and Research Center
NEIS	National Electrical Installation Standard
NETA	InterNational Electrical Testing Association
NFPA	National Fire Protection Association
NORM	Naturally Occurring Radioactive Material
NRC	Nuclear Regulatory Commission
NTSB	National Transportation Safety Board
OOS	Out of Service
OSBL	Outside Battery Limits
OSHA	Occupational Safety and Health Administration
P&ID	Piping and Instrumentation Diagram
PCB	Polychlorinated Biphenyls
PCM	Probability Centered Maintenance
PD	Partial Discharges
PFD	Probability of Failure on Demand
PHA	Process Hazard Analysis
PM	Preventive Maintenance
PS	Process Safety
PSB	Persistent Slip Bands
PSI	Process Safety Information
PSM	Process Safety Management
PT	Process Technology
PVC	Polyvinyl Chloride
QA	Quality Assurance
RAGAGEP	Recognized and Generally Accepted Good Engineering Practices
RBD	Risk Based Decisions
RBI	Risk Based Inspection
RBPS	Risk Based Process Safety
RCA	Root Cause Analysis
RCRA	Resource Conservation and Recovery Act
ROV	Remotely Operated Valves
ROW	Right of Way
RP	Recommended Practice
SAIDI	System Average Interruption Duration Index
SAIFI	System Average Interruption Frequency Index
SB	Service Bulletin

SCC	Stress Corrosion Cracking
SDOT	State Departments of Transportation
SIDD	Standard Installations Direct Design
SIF	Safety Instrumented Functions
SIL	Safety Integrity Level
SIS	Safety Instruments Systems
SMART	Specific Measurable Attainable Realistic Timely
SME	Subject Matter Expert
SSV	Safety Shutoff Valve
TENORM	Technologically Enhanced Naturally Occurring Radioactive Material
TIRP	Targeted Infrastructure Replacement Programs
TM	Thickness Measurements
UFC	Unified Facilities Criteria
UK	United Kingdom
UPS	Uninterruptible Power Supply
USA	United States of America
USDA	United States Department of Agriculture
VCS	Vapor Control System
WEF	Water Environment Federation
WFD	Widespread Fatigue Damage
WO	Work Order
WWT	Waste Water Treatment

REFERENCES

ACPA 2012, *Design Data M9: Standard Installations and Bedding Factors for the Indirect Design Method.* American Concrete Pipe Association.

API 2008. Recommended Practice 581, *API Recommended Practice 581 - Risk Based Inspection Technology.* American Petroleum Institute, Washington DC.

API 2011, Recommended Practice 571, *Damage Mechanisms Affecting Fixed Equipment in the Refining Industry.* American Petroleum Institute, Washington DC.

API 2007, Recommended Practice 579, *Fitness-for-Service.* American Petroleum Institute, Washington DC.

API 2009, Recommend Practice 580 (2009), *Risk-Based Inspection,* American Petroleum Institute, Washington DC

API 2014a, Recommend Practice 583, *Corrosion Under Insulation and Fireproofing,* American Petroleum Institute, Washington, D.C.

API 2014b, Recommend Practice 584, *Integrity Operating Windows,* American Petroleum Institute, Washington DC

ASCE (2010). *Recommendations for Design of Reinforced Concrete Pipe.* Ece Erdogmus, et. al., Journal of Pipeline Systems Engineering and Practice.

Baker, R. (2006). *Process Safety Concerns Can Arise When Using Refurbished or New-Surplus Equipment.* Hydrocarbon Processing Magazine, pp.73-80.

BPA (2014). *Asset Management Strategies.* Bonneville Power Administration, BPA Policy 240-2.

British Standard 7910 (2013). *Guide on methods for assessing the acceptability of flaws in metallic structures.* United Kingdom.

CCPS (1999). *Guidelines for Chemical Process Quantitative Risk Analysis.* Center for Chemical Process Safety of the American Institute of Chemical Engineers, New York.

CCPS (2001). *Layer of Protection Analysis: Simplified Process Risk Assessment.* Center for Chemical Process Safety of the American Institute of Chemical Engineers, New York.

CCPS (2005). *Building Process Safety Culture: Tools to Enhance Process Safety Performance.* ISBN # 0-8169-0999-7, Center for Chemical Process Safety of the American Institute of Chemical Engineers, New York.

CCPS (2006). *Guidelines for Asset Integrity Systems.* Center for Chemical Process Safety of the American Institute of Chemical Engineers, New York.

CCPS (2006). *The Business Case for Process Safety.* Center for Chemical Process Safety of the American Institute of Chemical Engineers, New York.

CCPS (2007). *Guidelines for Risk Based Process Safety.* Center for Chemical Process Safety of the American Institute of Chemical Engineers, New York.

CCPS (2008). *Inherently Safer Chemical Processes: A Life Cycle Approach,* Second Edition. Center for Chemical Process Safety of the American Institute of Chemical Engineers, New York.

CCPS (2009). *Guidelines for Process Safety Metrics.* Center for Chemical Process Safety of the American Institute of Chemical Engineers, New York.

CCPS (2016). *Guidelines for Safe Automation of Chemical Processes,* 2nd Edition Center for Chemical Process Safety of the American Institute of Chemical Engineers, New York.

CIDB. *National Infrastructure Maintenance Strategy (NIMS).* Infrastructure Maintenance Budgeting Guideline.

Clean Harbors (2016). *Hazardous Waste Transportation and Disposal.*

DCT 245 (2013). *Pavement Condition Surveys – Overview of Current Practices.* Delaware Center for Transportation.

DOD UFC 4-860-03 (2008). *Railroad Track Maintenance & Safety Standards.* Unified Facilities Criteria, US Department of Defense.

Dräger (2007). *Functional Safety and Gas Detection Systems Safety Integrity Level – SIL.* Dräger Safety AG & Co. KgaA, Germany.

Ebert, I., and Ochsenkuhn, H. (2011). *Ageing Infrastructure.* MUNICH RE, Schadenspiegel.

EC JRC (2013). *Corrosion-Related Accidents in Petroleum Refineries.* European Commission Joint Research Centre, Luxembourg:

EPA (2008). *Instrument Equipment Testing, Inspection and Maintenance.* Quality Assurance Handbook, Vol. II, Section 11.0.

Gage (2013). *Equipment Maintenance and Replacement - Decision Making Processes.* A Senior Project Submitted to the Faculty of California Polytechnic State University, San Luis Obispo.

Horrocks, P., Mansfield, D., Parker, K., Thompson, J., Atkinson, T., and Worsley, J. (2010). *Managing Ageing Plant, A Summary Guide.* Health and Safety Executive (HSE), United Kingdom.

Horrocks, P., Mansfield, D., Thompson, J., Parker, K., and Winter, P. (2010). *Plant Ageing Study – Phase 1 Report.* Health and Safety Executive (HSE), United Kingdom.

HSE (2006). *Plant Ageing, Management of Equipment Containing Hazardous Fluids or Pressure.* HSE Research Report RR509. Health and Safety Executive (HSE), United Kingdom.

HSE (2010). *Plant Ageing Study Phase 1 Report.* Health and Safety Executive Research Report RR823, Health and Safety Executive (HSE), United Kingdom.

HSE (2013). *E/C&I Plant Ageing: A Technical Guide for Specialists Managing Ageing E/C&I Plant.* Health and Safety Executive (HSE), United Kingdom.

IChemE (2013). *The Importance of Recognizing and Managing Ageing Plant.* Loss Prevention Bulletin 234, United Kingdom.

IEC 61511-1 (2016). *IEC 61511 - Functional Safety - Safety Instrumented Systems for the Process Industry Sector - Part 1: Framework, Definitions, System, Hardware and Application Programming Requirements.* International Electrotechnical Commission, Switzerland.

IEC 61511-2 (2003). *IEC 61511 - Functional Safety - Safety Instrumented Systems for the Process Industry Sector - Part 2.* International Electrotechnical Commission, Switzerland.

IEEE 400 (2012). *Guide for Field Testing and Evaluation of the Insulation of Shielded Power Cable Systems Rated 5 kV and Above.* Institute of Electrical and Electronic Engineers.

IEEE P1185 (2010). *Recommended Practice for Cable Installation in Generating Stations and Industrial Facilities.* Institute of Electrical and Electronic Engineers.

INGAA (2012). *Draft Work In Progress: Definition and Application of Fitness for Service to Gas Pipelines.* INGAA Integrity Management Continuous Improvement Work Group 4, Interstate Natural Gas Association of America.

Iseley, T. (1999) *Development of a New Sewer Scanning technology,* No-Dig America, May.

Kelly, B. (1998). *Integrated Risk Management - A Participative Approach.* Presentation to International Quality and Productivity Conference, Toronto, Ontario, September 28-30.

Kelly, B. (2004). *Private Transcript.* Risk Management Consultant.

Li, W., Vaahedi, E., and Choudhury, P. (2006). *Power System Equipment Aging.* IEEE Power and Energy Magazine, 4(3): 52-58.

Little, R. (2012). *Managing the Risk of Aging infrastructure.* International Risk Governance Council (IRGC), Public Sector Governance of Emerging Risks, November.

Megger (2006). A Stich in Time - *The Complete Guide to Electrical Insulation Testing. www.megger.com.*

Nimmo, W. and Hinds, G. (2003). *Beginners Guide to Corrosion.* National Physical Laboratory, Middlesex, United Kingdom.

NRC (2015). *Regulation of Radioactive Materials.* United State Nuclear Regulatory Commission.

NTSB (1989). *Aloha Airlines Accident Report AAR/89/03.* National Transportation Safety Board.

NY Times (1995). *Collapse of Apartment Building in Harlem Kills 3.* Richard Perez-Pena, The New York Times, March 22.

Pokluda, Jaroslav, and Andera (2010). *Micromechanisms of Fracture and Fatigue.* In a Multi-scale Context, Engineering Materials and Processes. *Springer Science + Business Media,* Berlin.

PSE&G (2015). *PSE&G Receives Approval of $905 Million Program to Accelerate Replacement of Aging Gas Infrastructure.* PSE&G Press Release, November.

Purdue University (2007). *Sewer Scanner and Evaluation Technology (SSET).* Division of Construction Engineering and Management.

Rodewald, J. (2014). *Six Proven Environmental Cleanup Methods.* Hazardous Waste Experts. http://www.hazardouswasteexperts.com/6-proven-environmental-cleanup-methods/

Sastry, V.P. (2015). *Corrosion Under Insulation and Fire Proofing Materials.* Seventh Middle East NDT Conference & Exhibition, Bahrain, September.

SCE (2015). *2015 General Rate Case, Transmission and Distribution (T&D).* Volume 4 – Infrastructure Replacement Programs, Southern California Edison, SCE-03, Vol. 04.

Schneider (2012). *Theoretical Review.* APC by Schneider Electric.

SD DOT (1995). *Rural Road Condition Survey Guide.* SD95-16-G1, SD Department of Transportation, Office of Research.

SLC (2013). *Audit and Inspection, Code 31F – Marine Oil Terminals, Section 3102F.* California State Lands Commission (SLC), Div. 2.

Stephens, G., and Stickles, R.P. (1992). *Prioritization of Safety Related Plant Modifications Using Cost-Risk Benefit Analysis.* International Conference on Hazard Identification and Risk Analysis, Orlando.

TCEQ (2015). *Radioactive-Waste Disposal: NORM Disposal.* Texas Commission on Environmental Quality (TCEQ), December.

US Chemical Safety Board, CSB (2016). *CSB Board Members Add Preventive Maintenance to Most Wanted Safety Improvement Program.* http://www.csb.gov/csb-board-members-add-preventive-maintenance-to-most-wanted-safety-improvement-program/

US DHS (2010). *Aging Infrastructure: Issues, Research, and Technology.* Buildings and Infrastructure Protection Series, Homeland Security Science and Technology, December.

USCG (2004). *Waterfront Facility Compliance Booklet USCG-5562A.* Department of Homeland Security.

USCG (2014). *Marine Safety Manual Volume II: Materiel Inspection.* COMDTINST 16000.7B, US Coast Guard.

Vaisnys P., Contri P., Rieg C., Bieth M., (2006). *Monitoring the Effectiveness of Maintenance Programs Through the Use of Performance Indicators.* EUR 22602 EN Report Netherlands.

Wasileski, R. (2014). *Retired & Dangerous Out-of-Service Equipment Hazards.* American Society of Safety Engineers (ASSE).

Wenyuan, L., Vaahedi, E., and Choudhury P. (2006). *Power System Equipment Aging.* IEEE Power & Energy Magazine, May/June.

Whitepaper. (2013). *The Impact of Aging Infrastructure in Process Manufacturing Industries.* The Economist, Intelligence Unit, Ltd.

Wintle, J., Moore, P., Henry, N., Smalley, S., and Amphlett, G. (2006). *Plant Ageing, Management of Equipment Containing Hazardous Fluids or Pressure.* HSE Research Report RR509. Health and Safety Executive (HSE), United Kingdom.

Wood, M.H., Vetere-Arellano, A.L., Van-Wijk, L. (2013). *Corrosion-Related Accidents in Petroleum Refineries.* JRC Scientific and Policy Report, EUR 26331 EN, European Union.

APPENDIX A

AGING ASSET CASE STUDIES

Figure AP.1-1. Example of an Old Facility Presenting Aging Signs

CASE STUDY 1: GAS DISTRIBUTION PIPELINE EXPLOSION

Example of ignoring warning signs for an asset that had exceeded its intended lifecycle.

An explosion and fire ripped through a residential neighborhood in Allentown, Pennsylvania on the evening of February, 2011 killing five people and injuring at least twelve others. The explosion destroyed eight homes and damaged fifty nearby businesses while forcing the evacuation of five hundred people.

The blast was caused by the failure of a 12-inch low pressure gas main, made of cast iron that had been installed in the 1920s. While no direct ignition source was determined, it was later learned that a 1979 work order recommended replacement of the line with 12-inch steel after four breaks were experienced in the period 1974 to 1979. The line was never replaced and the work order was misplaced for several years.

In subsequent years, two other fatal gas explosions were experienced in Allentown as a result of cast iron gas main breaks. Again, several people were injured and property damage was extensive. Separate requests were made for the local gas company to replace the mains. A September 1979 gas company inspection discovered a moderate leak in a gas main but concluded that it posed no immediate danger to the public. However, it was admitted by the company that such a leak could grow over time. The maintenance strategy of the gas company has since been challenged.

CASE STUDY 2: MISSISSIPPI BRIDGE COLLAPSE

An eight- lane steel truss bridge collapsed in Minneapolis, Minnesota during the evening rush hour in August 2007 killing thirteen people and injuring 145 others. The bridge, one of the busiest in the state, was built in 1967 and had 14 sections spanning nearly 2000 feet over the Mississippi River. It carried eight lanes of traffic with no apparent load or traffic restrictions. The bridge was comprised mainly of steel trusses and deck plate. In the early planning stages heavy industry was forced to vacate the region in close proximity to the proposed bridge ramps. Toxic waste and contaminated soil were later discovered and a major clean-up effort was required before construction could begin. It is not known how effective this clean-up initiative was. The relationship between this fact and the eventual collapse has not been established but is still questioned by many authorities.

In December 1985, a major traffic pile up occurred on the same bridge as a result of black ice formation. The problem was compounded by the smooth surface of the driving lanes and close proximity of a waterfall that dispersed water droplets into the atmosphere. In later years, the Department of Transportation began r experimenting with magnesium chloride solution and corn-processing byproduct to reduce black ice formation. Finally, in 2000 a system of temperature-activated nozzles was installed to spray potassium acetate solution onto the bridge deck in order to keep the area ice free.

Regarding the condition of the bridge, as early as 1990 federal government inspections revealed serious engineering and corrosion concerns within the bridge. Problems with cracking and fatigue were noted and a plan was announced that the bridge would be replaced in 2020. In the period leading up to the collapse extensive repairs were recommended to the bridge. These were cancelled in favor of periodic safety inspections. It was feared that any repair work might further weaken the bridge structure. It is worth noting that at the time of the collapse, four of the eight lanes were closed for resurfacing, and there were 575,000 pounds (261,000 kg) of construction supplies and equipment on the bridge.

Official investigation reports cite this tragedy as an engineering failure. Simply put, the original design failed to consider the load conditions and environment to which the bridge would be exposed. Nonetheless, aging was a significant contributing factor aggravated by the thoughtless application of de-icing salt.

CASE STUDY 3: SINKING BUILDING FOUNDATION

Structural foundation damage can result in serious consequences, such as collapse of a structure with associated casualties. This case study is not specific to any incident. Nonetheless, foundation failures resulting in deaths from the causal mechanisms described below have occurred frequently in developing countries. Fact is, such events can occur anywhere if soil conditions are not monitored and addressed.

Structural damage can occur for several reasons, such as improper preparation of the subsoil when a building's foundation is originally excavated. Most commonly, though, foundations sink or drop because of expansion and subsequent contraction of the soil that supports them. This becomes a more frequent occurrence in periods of ongoing drought.

Although different types of soil react differently to moisture extremes, the clay soil common in the US Midwest is what is known as expansive soil, that is, it swells when saturated with water. This is of particular concern in homes where exterior water management is not practiced by keeping gutters clean and extending downspouts away from the house, causing the soil around and under the foundation to expand during wet periods. When extended dry conditions occur, this over-saturated soil shrinks, causing the foundation footings to drop and the foundation walls, along with parts of the above-ground structure, to crack and separate.

It is usually pretty easy to tell when a foundation has sunk. Early signs of this type of structural foundation damage include jammed windows and doors and drywall cracks in the above-ground living space. More advanced or severe damage is indicated by wide cracks in exterior walls or even separation of one section of a building from the rest, particularly common in additions.

Dropped or sunken foundations can only be repaired by underpinning, a process in which a support pier is placed under the foundation, lifting it back up to level and stabilizing it. There are several types of piers, including concrete and helical piers, but the preferred method is to use a hydraulically driven steel pier, sometimes called a "push" pier.

What is important to know about the timing of such repairs is that the damage is usually progressive, starting out relatively small and worsening as the soil under the foundation continues to shrink and the foundation continues to drop. The good news is that even a severely damaged foundation can still be underpinned and restored to level. However, there are bad news sometimes associated with a damaged foundation. The worse the damage gets, the more piers are required to be repaired and, logically, the cost of the repair increases with each pier that is installed. A sunken corner that is detected and repaired early might be fixed with four or five piers. However, an entire wing or large section of a building that drops gradually over time, might require many piers at significant cost and with questionable benefit.

CASE STUDY 4: TAILINGS DAM FAILURE

In April 1998, a tailings dam failed at a lead-zinc mine in southern Spain, releasing nearly 5 million cubic meters of toxic slurry and liquid into a nearby river. The slurry wave covered several thousand hectares of farmland and resulted in extensive environmental damage. Mining had commenced in the late

80s and the earth dam was constructed to contain acidic effluent slurry from the ore extraction process. The dam was constructed on carbonate rich bedrock and at the time of the failure its height had reached 25 meters. While improper design of the dam was a major contributor, the investigation indicated that the breach was caused from chemical attack of the impounded acidic pyritic slurries on the rock forming the dam foundation material. A consultant report two years earlier had identified a potential weakness in the dam structure as a result of ground movement within the foundation but no action had been taken. The cleanup effort required 3 years.

This serious incident serves to illustrate that mining structures such as dams are in a continual state of flux due to the forces of gravity and potential fluid seepage. Extensive monitoring is required along with a commitment to take action when circumstances warrant it.

CASE STUDY 5: SINKING OF THE BETELGEUSE

Betelgeuse was a first-generation oil supertanker built in 1968 and operated by Total SA. Having a rated capacity of 121,000 tons (DWT) it was used to ship Middle East crude to refineries in western Europe. However, towards the end of the seventies, a new generation of larger supertankers was commissioned and this threatened the future of the Betelgeuse. Accordingly, regular maintenance was reduced or deferred as the end of its lifecycle approached. It was reported that the ship had experienced severe internal and external corrosion including the bulkheads, which isolated the various oil compartments.

In late 1978 the Betelgeuse was dispatched to deliver a shipment of crude oil to Portugal. As a result of logistics issues and severe weather in central Europe the ship was diverted to Whiddy deep water terminal in western Ireland. En route to the terminal the ship sustained some minor damage causing oil to leak from the bulkheads, a possible warning sign of future problems. The origin of the leaks was determined and temporary repairs were made.

Upon arriving at Whiddy terminal the Betelgeuse was anchored adjacent to a loading rack on a concrete jetty, 1300 ft from a large onshore tank farm. The terminal had been constructed a decade earlier but due to neglect its firewater systems had deteriorated and chronic leakage problems became common. To reduce operating costs the firewater systems were placed in standby mode only to be commissioned when needed.

Oil transfer operations commenced in the early morning hours of January 8, 1979. A sequential unloading procedure was used to ensure that oil was evenly removed from each of the storage compartments thereby reducing bending stresses on the hull of the aging ship (shown as up thrust arrow for empty compartments and down thrust arrow for full compartments). It was not determined whether this procedure was properly followed on that day, since there were no survivors from the incident that followed. With little warning the hull began to crack and loud noises were heard followed by fires on the upper deck. The crew started to abandon ship and muster on the jetty. With no firewater available the crew anxiously awaited rescue from an emergency vessel. Before the emergency rescue boat arrived a series of violent explosions erupted causing the hull of the ship to break in half. Fifty people were killed including crew members and rescue workers. This event serves to illustrate the importance of maintaining all critical facilities and infrastructure in reliable condition and a state of readiness.

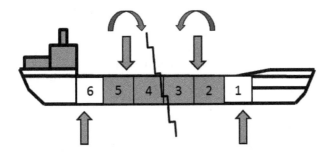

Figure AP.1-2. Sketch Showing Bending Moment as a Result of Unbalanced Buoyancy Forces

CASE STUDY 6: ALEXANDER KIELLAND DRILLING RIG DISASTER

In March 1980, a large oil rig capsized in the North Sea killing 123 workers. The Alexander Kielland was a semi-submersible drilling rig that was constructed in 1976. Following initial commissioning it was modified to provide offshore housing for workers from the Ekofisk oil field off the coast of Norway. There were 212 workers on board, one stormy evening, when a large wave hit the rig rupturing one of its five support legs. This caused most of the anchor cables to snap creating unbalanced stresses on the other legs. Within the next fifteen minutes several workers managed to escape using lifeboats or by diving directly into the icy waters below. When the final cable snapped, the rig capsized killing all remaining workers.

Following extensive salvage operations, parts of the wreckage were towed back to Norway for analysis. It was later determined that the failed leg had a fatigue crack in it which had existed prior to the accident but had not been detected. The origin of the crack was traced to a small fillet weld which joined a non-load-bearing flange plate to the main bracing. The poor-quality fillet weld contributed to a reduction in the strength of the support leg. Cold cracks in the welds, increased stress concentrations due to the weakened flange plate, and cyclical stresses (resulting from North Sea exposure) all contributed to the rig's ultimate collapse. The failure that led to this tragedy was inevitable since there were no inspection initiatives in place that might have identified the cracks.

CASE STUDY 7: ROOF COLLAPSE AT ORE PROCESSING FACILITY

A large ore processing facility located in the northwest was built in the late 70s. Because the operation involved steam and water and it was exposed to colder weather, parts of the operation were enclosed in a large building. From a distance the building appeared similar to an aircraft hangar. The 60-foot-high roof was constructed of steel panels supported on open web steel joists similar to that of a large warehouse. Over a period of 20 years, steam and high humidity within the building caused water to condense on the inner surface of the roof panels resulting in severe rusting of the joists. Occasionally, water condensate and rust fragments dropped to the annex below but these symptoms were ignored since they were not deemed to be process related.

One winter evening, following a heavy snowfall, a large section of the roof collapsed onto the floor below. Extensive equipment damage resulted but fortunately there were no injuries owing to the occurrence during evening hours. A formal investigation revealed cyclic melting and freezing of snow causing a heavy layer of ice to form upon the roof. The roof joists had totally disintegrated from rust. In essence, the ice was actually supporting the roof prior to failure.

This incident could have had tragic consequences. The failure was certainly age related and was the result of prolonged conditions that had not been considered or addressed during the design some 20 years earlier. Furthermore, the pending failure was not fully visible from the floor level and no party was directly responsible for inspecting the roof.

INDEX